HURRY UP SLOWLY
Stay True To Yourself When You Navigate The Corporate World
And Get Rewarded Without Working Excessive Hours

ISBN: 978-1-947276-47-5
ASIN: B0C9F55XYN

Publishing and Design:

EPIC AUTHOR
PUBLISHING

Ordering Information: Exclusive discounts for quantity purchases.

Contact: 561-601-9871 | info@epicauthor.com | EpicAuthor.com

First Edition

HURRY UP SLOWLY

Stay true to yourself when you navigate the corporate world and get rewarded without working excessive hours

SURAJ ETHIRAJAN

This book is dedicated to you.

May you amplify your voice and find safety to thrive.
I am wishing for your leadership to shine brighter. I applaud you for taking steps to leave an enduring mark on this world and improve it for future generations.

—Suraj Ethirajan

Table Of Contents

Praise for Hurry Up Slowly

by **Jack Dunnigan**,
Sales and Business executive in Silicon Valley, CA, USA

As a seasoned business and sales executive, I've had the privilege to lead teams, define product strategies, and generate multibillion-dollar revenues. Amidst the hectic pace of the corporate world, finding a resource that empowers you to navigate challenges and attain ongoing success is rare. I'm ecstatic to introduce you to *Hurry Up Slowly* by Suraj Ethirajan.

I've known Suraj for years, and his wisdom, exuberance, and pragmatic approach have consistently impressed me. And with *Hurry Up Slowly*, he has outdone himself. This book speaks directly to the challenges we face in today's rapid corporate world—whether you're a tech professional seeking work-life balance, an engineer yearning to reignite your passion, or a senior leader looking to scale your impact.

Suraj shares practical techniques to regain control, enhance productivity, and unlock potential. I found his insights on time management and embracing mindfulness particularly powerful. His blend of wisdom and unfiltered energy creates a compelling narrative that inspires and equips you to take action.

What you are about to read is more than a book. It's a transformational expedition towards a more fulfilling and triumphant life. I've personally witnessed the impact of Suraj's coaching services. He helps you embrace intentional living, become the architect of your success story, and, most importantly, thrive in your career journey.

Hurry Up Slowly is a route to success. If you want to make noteworthy strides in your career and personal development, seize a copy and let Suraj coach you. Take the next step.

Introduction

Have you ever stood at the foot of a bridge and wondered what lies on the other side? That's the beauty of bridges—they link places, they link people, and, in a broader sense, they link ideas. This book is your bridge. It's a lifeline, connecting us in our shared journey.

But the beauty of a bridge lies not just in its ability to connect but also in its promise of a destination. Before crossing a bridge, you must understand where you stand. Once you know your starting point, it's only then that you can truly appreciate the journey to the other side.

In these pages, we'll embark on a conversation. While there will be frequent references to technology, or "tech," professionals, know this—I'm speaking to you. Yes, you. We'll explore your current situation, your "before," and identify the known and unknown apprehensions that can hold you back.

From there, we'll envision your "after," the career growth you aspire to achieve. The scenarios are many, but the choice is yours. Pick what resonates with you and what applies to your unique journey. Remember, you're the architect of your life; this book is your blueprint. So, shall we begin?

BRIDGING THE GAP: A SNEAK PEEK INTO YOUR "BEFORE" AND "AFTER" STATES IN TECH

The Before

Have you ever taken a moment to consider what might be holding you back on this side of the bridge? What worries and uncertainties do tech professionals grapple with when no one else is watching?

One of the most common fears is job security. In an increasingly digital world, the specter of job redundancy looms large. This worry was further amplified by the layoffs of over 200,000 tech workers in 2023 and another 154,000 the year prior due to shifts in consumer spending and rising interest rates.

Burnout is another significant concern. It often sprouts from poor leadership, excessive workloads, a toxic work culture, and limited opportunities for career growth. Many tech professionals aspire to continuous learning and advancement but are hampered by a lack of development opportunities.

Moreover, the hiring process can also be a source of apprehension. Despite research suggesting that people can learn new skills, unconventional tech hires often face skepticism.

But there's more. Beneath the surface, there are quieter, less recognized apprehensions. The constant need to adapt and learn new tools and systems can result in stress and burnout. Social media usage can induce a fear of missing out, negatively impacting mental well-being and focus.

Furthermore, navigating personal circumstances while maintaining a demanding tech career can be challenging. Life events such as parenthood, caring for aging parents, or coping with personal health issues can add to the stress.

And then there's the danger of complacency—the risk of becoming like the proverbial frog in boiling water, oblivious to the gradually increasing heat until it's too late. In the rapidly evolving tech landscape, there's a real risk of overlooking the need for continuous improvement and becoming obsolete.

These apprehensions, whether clearly acknowledged or subtly felt, can lead to stress, decreased productivity, and lower job satisfaction. They can trigger a range of emotions—from anxiety due to the pace of technological change to feelings of devaluation and disillusionment, dissatisfaction with work routines, and uncertainty about market changes.

Upon recognizing these apprehensions, reactions can vary. Tech professionals may experience surprise, anxiety, curiosity, determination, denial, or resilience. Despite the challenges, many show an extraordinary capacity to devise coping strategies, persist, and ultimately thrive in this ever-evolving tech landscape.

The After

For the tech professional, a successful career vision is not merely a dream but an achievable reality. This envisioned future is one of job security, growth, stability, and a work culture that appreciates individual contributions while providing clear paths for career progression.

Imagine a world where work and personal life coexist harmoniously. The workload is manageable, and personal time is respected. In this scenario, continuous learning and development are at the heart of the journey. Adapting to new systems and tools at a comfortable pace becomes an integral part of professional growth.

Envision a thriving work culture where strong leadership is valued and a sense of community is fostered. Here, unconventional skills and diverse backgrounds are celebrated. Staying ahead of the curve becomes the norm, job satisfaction is high, and each individual's work significantly contributes to the company's success.

Visualize a landscape abundant with growth opportunities. Rapid changes in the tech industry are met with learning agility, ensuring adaptability and relevance. The growth of the tech industry provides job security and numerous opportunities for career advancement.

In this ideal state, flexible work hours and options are standard practices. Innovative employers become daily collaborators. High-paying jobs are the norm, and traits like a strong work ethic, curiosity, and effective communication are highly valued.

Mastering the art of negotiation leads to better pay, exciting projects, and improved working relationships. This ideal state, where wants and needs are exceeded, is not a distant dream. It's a possibility within reach, subtly beckoning every ambitious tech professional to seize it. This fulfilling and successful career could be just around the corner, waiting to be discovered and embraced.

The Bridge

Imagine yourself as the protagonist in an epic movie. Your life is the plot, your challenges are the conflicts, and this book, *Hurry Up Slowly*, is the bridge that leads you to your triumphant resolution.

In the opening scenes, you'll embark on a journey of self-discovery. This book will serve as your mirror, reflecting your purpose, knowledge, and values-based leadership. You'll understand the depth of your character, setting the stage for the transformation to come.

As the plot unfolds, you'll encounter the conflicts of your story—your circumstances, challenges, and the battles you wage in the tech workplace. But fear not, for these are not roadblocks but springboards to a higher plane. With the tools and techniques shared in this book, you'll learn to leap over these hurdles with grace and resilience.

The climax of your narrative arrives when you aim for your North Star, your ultimate professional goal. This book will be your guide, teaching you unique skills and enhancing your storytelling abilities. As you narrate your journey, you inspire others, leaving an indelible mark on your world.

As the story starts to resolve, you'll discover how consistency compounds, ways to build effective habits, and the power of teaching others. These lessons will be your roadmap to personal growth and long-term success, turning the ordinary into extraordinary.

Finally, in the grand finale, you'll be guided to reevaluate your values and reinvent yourself periodically. This is the secret to staying relevant and adaptable in an ever-changing tech landscape. Empowered with the wisdom of lifelong learning, you'll sail forward, ready to face new challenges and seize new opportunities.

This journey, dear reader, is your story. It's about becoming the best version of yourself and reaching your full potential as a tech professional. Embark on this journey with *Hurry Up Slowly* and watch your transformation unfold, scene by inspiring scene.

THE ART OF HURRYING UP SLOWLY: A DEEP DIVE INTO THE ACTIVITIES THAT SHAPE YOUR JOURNEY

In the grand narrative of your journey, "Hurry Up Slowly" is not just the title of this book. It's the underlying philosophy, a pearl of paradoxical wisdom that drives every page, every story, and every activity within these covers.

Think of it as a director's advice to the lead actor—you. The wisdom, tools, advice, and stories are your script, but the activities are your rehearsals. They are thoughtfully engineered to provide hands-on experiences and growth opportunities at every stage of your journey.

These activities, dear reader, are your "hurry up" moments. They push you to act, write, capture thoughts, survey your trusted circle of friends and family, review, redo, and understand deeper. They demand your engagement, your energy, and your time. They push you out of your comfort zone, challenging you to grow, learn, and improve.

However, don't forget the "slowly" part amid this hustle. Take your time to reflect, understand, and internalize what you learn. Writing isn't just about putting thoughts on paper; it's about making sense of those thoughts, giving them shape and direction. It's about painting the canvas of your reality with careful, thoughtful strokes.

The activities might seem challenging, even daunting, but remember, growth often comes from discomfort. It's the hard work, the sweat, and the determination that brings success. As you engage with these activities, you're not just going through tasks. You are laying the foundation of your future, brick by brick.

So, embrace this philosophy of "Hurry Up Slowly." Hurry up to act, challenge yourself, and step out of your comfort zone. But do it slowly, with patience and perseverance, knowing that each step brings you closer to your goals, no matter how small.

CHAPTER 1

Setting The Scene For Your Career Development – The Exposition

Welcome to the first chapter of a transformative journey that promises to unveil the extraordinary within you. Prepare to step into a world where your purpose becomes your compass, values are your guiding light, and knowledge is the key to infinite possibilities.

Ever wondered why some people seem to have an unwavering drive, an undeniable passion that fuels their actions? It's as if they've discovered a secret formula for success. Well, the secret is out, and it begins with discovering your purpose. Dive into this section and unlock the power of "why." Find your unique calling and let it illuminate your path to success.

Do you aspire to be a leader who inspires, motivates, and creates an environment of mutual trust and respect? This section unravels the concept of values-based leadership. Discover how aligning your actions with your core values can transform you into a leader that others not only respect but aspire to emulate.

In an ever-evolving world, knowledge is your superpower. But it's not just about acquiring knowledge; it's about learning, unlearning, and relearning. This section will guide you through the labyrinth of continuous learning, empowering you to stay ahead of the curve and navigate the waves of change with ease.

This chapter holds the promise of transformation, growth, and success. As you turn each page, you'll uncover strategies and insights that will propel you toward your desired outcomes. Are you ready to embark on this exciting journey? The road to your best self awaits. Let's dive in!

SECTION 1: PURPOSE

What is it that propels you forward, stirs your soul, and breathes life into your days? Have you found your purpose, or are you still seeking it?

The quest for purpose is as old as humanity itself. It's an intricate dance of self-discovery and introspection, of understanding our desires and aligning them with our actions. As we embark on this journey, we often find ourselves standing at the intersection of passion and confusion, wondering which path to take.

Imagine you are walking down a bustling street, lost in the crowd. Suddenly, someone taps on your shoulder and asks, "What fulfills you?" How would you respond? Would you be able to articulate your passions and aspirations, or would you find yourself grappling for words?

Our purpose isn't just a destination; it's a compass that guides us through the labyrinth of life. It's a force that pushes us to scale heights, even when the world around us is pulling us down. But how do we ensure that we stay true to our purpose, especially when the winds of adversity blow hard?

The answer is both simple and profound—write it down. When we put pen to paper, our purpose comes alive. It morphs from a mere idea floating in our minds to a tangible entity that we can see, touch, and feel. It becomes a beacon of light, guiding us through the darkest of nights and the stormiest of seas.

Our lives are a whirlwind of tasks and responsibilities. We juggle priorities, race against time, and strive to strike a balance between our professional and personal lives. Amidst this chaos, however, there's another dimension that's often overlooked—whether the activity resonates with our purpose and contributes to our personal growth.

This reminds me of the story of the old woodsman and his axe. The woodsman worked tirelessly, day in and day out, chopping down trees with his trusty old axe. No matter how hard he worked, though, he could never chop down more than a handful of trees a day. One day, a young man approached him and said, "Sir, if you take some time to sharpen your axe, you will be able to chop down many more trees."

But the woodsman replied, "I don't have time to sharpen my axe. I'm too busy chopping trees."

Like the woodsman, we too often get so caught up in our daily tasks that we forget to sharpen our axes—to align our activities with our purpose and ensure they contribute to our growth. By doing so, we not only enhance our productivity but also add meaning to our lives.

How do you discover your passion in the realm of technology? Can it be born out of the era or environment you find yourself in, or is it something more intrinsic?

As you embark on this journey through the pages of this book, remember that it's more than just a reading experience. It is an invitation to self-discovery. This book, with its worksheets and activities, is designed to be your personal journal—a tangible record of your thoughts, commitments, and progress.

I encourage you to pick up a pen or pencil and actively participate in each exercise. Write down your reflections in a notebook specifically while you read and hurry up slowly. Don't just skim through the questions. Pause, ponder, and pen down your responses. Remember, there are no right or wrong answers, only insights that will guide you toward a fulfilling life.

The initial two chapters—the exposition and the rising action—focus on the "what" and the "why." What is your purpose? Why does it matter? We delve into these existential questions, helping you uncover your passions and align them with your actions.

These chapters are less about setting goals and more about understanding your purpose. They're about recognizing what stirs your soul and why it matters. This is a journey of discovery, not a race to the finish line.

But as we progress through the story, the focus shifts. In Chapters 3, 4, and 5, we turn our attention to the "how" and the "when." How can you build new skills? When should you multiply them? How can you continuously improve yourself? These chapters are about action and applying your purpose in real-world scenarios.

Just like a potter shapes clay into a beautiful piece of pottery, these chapters will help you mold your purpose into a meaningful career and a fulfilling life.

So, dear reader, are you ready to embark on this journey? Are you prepared to dive deep into the ocean of self-discovery and uncover the pearls of your purpose? If yes, then let's begin. Pick up your pen, open your mind, and let the journey commence.

Here's to discovering our purpose, to living our passion, and to making every moment count.

ACTIVITY: Discover Your Purpose – A Journey of Self-Reflection

In the hustle and bustle of life, we often lose sight of our purpose. This exercise is designed to help you reconnect with your inner compass and rediscover your purpose, both in work and life.

Materials Needed:

- A quiet space for reflection
- A pen or pencil
- A notepad or a journal

Instructions:

- **Find a Quiet Space:** Find a quiet and comfortable spot where you can sit undisturbed for a while. This could be a cozy corner in your home, a tranquil spot in a park, or even a quiet café.
- **Warm-Up:** Close your eyes and take a few deep, calming breaths. Let go of any tension or stress. Clear your mind, and prepare yourself for this journey of self-discovery.
- **Visualize:** Now, imagine your ideal day at work and in life. What does it look like? Who are you working with? What tasks are you performing? How do you feel at the end of the day?
- **Identify Your Purpose Words:** Open your eyes and write down words or phrases that describe your purpose. These could be verbs that represent actions you love doing (like "creating," "teaching," or "healing"), nouns that depict your passions (like "art," "technology," or "nature"), or adjectives

that describe the feelings you want to experience (like "fulfilled," "inspired," or "empowered").

- **Categorize:** Now, divide these words into two categories: work and life. Some words might fit into both categories, and that's perfectly okay.
- **Reflect:** Look at your words. Do they resonate with your current life and work? Do they align with your actions and decisions? If not, what changes do you need to make?
- **Plan:** Write down three steps you can take in the next month to align your life and work with these purpose words.
- **Revisit:** Make a commitment to revisit this exercise every few months. As you evolve, your purpose might evolve too. Regular self-reflection will ensure that you stay aligned with your purpose.

This is not a one-time activity; it's a journey of self-discovery. So, take your time, enjoy the process, and remember there are no right or wrong answers, only insights that will guide you toward a fulfilling life.

Remember to jot down your responses in your notebook or journal, creating a personal roadmap toward your purpose. This will serve as a physical reminder of your journey and progress. Happy discovering!

A Grandmother's Wisdom: Discovering Your Life Purpose

The summer sun was high in the sky, its rays dancing on the azure waters of the Indian Ocean. I found myself walking along the sandy shores with my grandmother, a woman who had been a presence in my life since the tumultuous days following the Gulf War in 1990.

As a young boy of eight, I remember being uprooted from our home in Kuwait and seeking refuge in India with my grandparents. Back then, I wasn't particularly fond of my grandmother. Our differences seemed too vast, and the cultural shift was challenging for me to comprehend.

However, life has a peculiar way of weaving together a tapestry of experiences that shape us. The turning point came when my grandfather passed away. Circumstantially, I was the only person with my grandmother at that moment. It felt as if the universe conspired to bridge the gap between us. Our relationship blossomed from that point forward, transforming into a bond of mutual respect and deep affection.

Breaking the comfortable silence between us, my grandmother turned to me, her eyes twinkling with curiosity. "What is it that you love to do? What is that one thing you feel qualified to teach others?" she asked, her voice a familiar melody of warmth and wisdom.

I pondered over her question, reflecting on my passions and skills. After a moment, I shared my thoughts with her. She nodded, her smile encouraging me to delve deeper into my self-discovery journey.

Next, she asked, "Who do you do it for?" I thought about my audience, the people I intended to serve through my passion. I voiced my answer, and she nodded again, her smile deepening.

Then came the third question, "What do they want or need from you?" I reflected on their needs, their desires, and what they sought from my service. I shared my thoughts, and she listened, her gaze fixed on the horizon.

The fourth question was profound. "How do they change or transform as a result of what you give them?" I contemplated the impact of my work and the transformation it could bring about. As I voiced my answer, I saw a spark of approval in her eyes.

Finally, she asked me to combine my answers into a sentence that answered five key questions:

1. Who am I?
2. What do I do?
3. Who do I do it for?
4. What do they want or need?
5. How do they change as a result?

As I recited my sentence, I felt a wave of realization wash over me. I had discovered my life purpose.

My grandmother turned to me, her eyes filled with wisdom and kindness. "You see," she said, "the happiest people are those who serve others and focus on making others happy. If you make other people happy, life will take care of you."

Before I left for the United States to pursue my undergraduate studies at ASU (Arizona State University), I saw my grandmother one last time. She shared a piece of advice that I carry with me to today—to turn my life purpose into a personal elevator pitch, answering the question, "So, what do you do?" with the transformational impact of my work.

Now, as I looked at the setting sun painting the sky in hues of orange and red, I realized that this was not just a walk on the beach. It was a profound journey of self-discovery, a lesson in finding my life purpose, and a testament to an unlikely bond that blossomed between a grandmother and her grandson.

ACTIVITY: Discover Your Life Purpose – A Self-Exploration Exercise

Welcome to the most exciting journey you will ever embark on—the journey of self-discovery. This exercise is designed to help you uncover your life's purpose, using a series of introspective questions and the insights of your trusted circle. Are you ready? Let's begin!

Materials Needed:

- A dedicated notebook or journal (this could be part of your *Hurry Up Slowly* activity book)
- Pens, colored pencils, markers, etc.
- Magazines, newspapers, or printouts for collage materials
- Glue
- A quiet space
- An open mind

STEP 1: Self-Reflection

Spend a week reflecting on the following questions. Write your answers in your journal. Don't rush this process. Take your time to consider each question deeply.

- Who am I?
- What do I do?
- Who do I do it for?
- What do they want or need?
- How do they change as a result?

Don't limit yourself to words alone. Use shapes, pictures, doodles—anything that helps you express your thoughts more vividly. You can also make collages from magazine cutouts to represent your answers visually.

STEP 2: Seek External Perspectives

Reach out to a trusted circle of family, friends, and colleagues. Share the same five questions with them and ask them how they perceive you in relation to these questions.

Remember, this isn't about seeking validation but gathering different perspectives to broaden your understanding of yourself. Note down their responses in your journal.

STEP 3: Synthesis and Reflection

With all the information at hand, spend another week to synthesize and reflect. Look for common themes, surprising insights, and areas of alignment or misalignment between your self-perception and how others perceive you.

Again, use words, shapes, pictures, and other creative means to capture your reflections.

STEP 4: Craft Your Life's Purpose Statement

Based on your introspection and the insights gathered, craft a sentence that encapsulates your life's purpose. This is not set in stone and can evolve as you grow and learn more about yourself.

Remember, this is a journey of self-discovery, not a race. So, hurry up slowly. Take your time to delve deep, understand yourself better, and discover your life's purpose. Enjoy this process, and embrace what you learn about yourself along the way.

"Courage is more exhilarating than fear, and in the long run, it is easier. We do not have to become heroes overnight. Just one step at a time."
—Eleanor Roosevelt

Exploring Values Related to Your Life's Purpose

We embark on another journey of self-discovery as we delve into the heart of what truly matters to us—our values. Picture your life as a wheel with seven spokes, each representing an essential aspect of your existence that directly influences your well-being and sense of accomplishment.

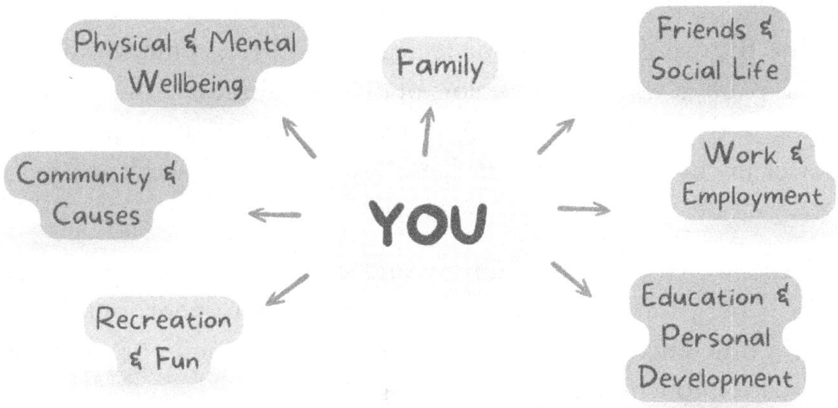

The first spoke represents the family. This could mean your birth family or the one you've chosen. Family is our anchor, offering us a safe harbor in the stormy seas of life. They shape our initial understanding of the world and often influence our core principles.

Now, let's move to the second spoke—friendship and social life. Friends, they say, are the family we choose. For some, this circle may occupy a larger segment of our daily interactions. As our world has shifted due to events like the pandemic, many of us found new ways to connect and engage with our friends, proving that physical distance can't dampen the bonds of friendship.

The third spoke symbolizes work and employment. Many of us spend a considerable portion of our lives at work. It's not just about earning a livelihood. Our jobs often reflect our passions, interests, and, most importantly, our purpose in life.

Let's turn our attention to the fourth spoke—education and personal development. Remember the exhilaration of learning something new during our school days? That joy doesn't have to end with formal education. The difference between the good and the great often lies in their commitment to lifelong learning, just like you are demonstrating by taking this course.

The fifth spoke is all about recreation and fun. Our world is a canvas painted with vibrant colors of diverse cultures, cuisines, landscapes, and entertainment. These recreational activities don't just offer us joy but also enrich our lives, influencing our work and personal well-being.

Next, we come to community and causes, the sixth spoke. As social beings, we yearn to be part of a community, contribute, and make a difference. Our purpose often extends beyond our personal boundaries, touching lives across the globe.

Finally, the seventh spoke represents physical and mental well-being. The importance of this spoke has been magnified in recent times with more emphasis on mental health than ever before. Just as you would put on your oxygen mask first in an airplane emergency, taking care of your mental health allows you to be there for others.

We'll delve deeper into each of these seven aspects, weaving them together with stories and experiences that will hopefully resonate with you. So, buckle up and get ready to explore the landscape of your values as we set forth on this enlightening journey of finding your purpose.

1. **Family:** Dr. Fe del Mundo,[1] a Filipino pediatrician, was driven by her family's experiences. After losing siblings to health conditions, she dedicated her life to advancing pediatric medicine in the Philippines. Her family-oriented values led her to

establish the first pediatric hospital in the Philippines. Born in 1911 in Manila, Philippines, Fe was the youngest among eight children. The death of her older sister due to appendicitis ignited a spark in young Fe to become a doctor. Despite societal norms and the patriarchal culture, she pursued medicine and graduated as valedictorian from the University of the Philippines College of Medicine. She later established the first pediatric hospital in her home country, transforming countless lives.

Three Takeaways:

- Personal experiences can fuel your passion.
- Transform personal loss into a force for change.
- Family values can shape your career.

2. **Friendship and Social Life:** Akio Morita and Masaru Ibuka,[2] cofounders of Sony, started with a shared vision and strong friendship. Their camaraderie and mutual respect fueled Sony's growth into a global tech leader. In post-World War II Japan, the two friends shared a vision to create a company that would be a leader in technology. With only a small amount of capital and a rented department store, they started Sony. Their friendship stood the test of time, and they remained partners until Ibuka's retirement. Today, Sony is a testament to their enduring friendship and shared vision.

Three Takeaways:

- A shared vision can strengthen friendships.
- Mutual respect is key to successful partnerships.
- Friendships can foster innovative ideas.

3. **Career and Employment:** Radia Perlman,[3] often referred to as the "Mother of the Internet," revolutionized network rout-

ing with her invention of the Spanning-Tree Protocol. Her dedication to problem-solving made networks more robust and efficient. As a young girl, Radia loved math but felt isolated because of her interest in what was then considered a male-dominated field. However, she didn't let societal norms deter her. She pursued her passion at the Massachusetts Institute of Technology (MIT), where she developed the Spanning-Tree Protocol. Radia proved that one's passion should never be hindered by societal expectations.

Three Takeaways:

- Curiosity and problem-solving can drive career success.
- Women are pioneers in technology.
- Patience and persistence pay off.

4. **Education and Personal Development:** Mary Jackson[4] fought for her right to education and broke racial barriers. Her thirst for knowledge led her to develop wind tunnel experiments that helped advance US space exploration. Mary grew up in Hampton, Virginia, during the segregation era. Despite the racial barriers, she excelled academically and earned degrees in mathematics and physical science. When NASA's Langley Memorial Aeronautical Laboratory began hiring black women, Mary seized the opportunity. She fought for her right to attend a whites-only school to gain the qualifications necessary to become an engineer.[4] Her thirst for knowledge and determination to break racial barriers made her NASA's first black female engineer.

Three Takeaways:

- Education can break barriers.
- Lifelong learning propels personal development.
- Fight for your right to education.

5. **Recreation and Fun:** Brewster Kahle,[5] a digital librarian, turned his hobby into the Internet Archive, a free digital library. His love for collecting and preserving digital content made millions of books, movies, and music accessible worldwide. As a child, Brewster was fascinated by libraries and the wealth of knowledge they offered. He dreamed of having access to a universal library at his fingertips. This childhood fascination turned into a lifelong passion, leading him to create the Internet Archive. Today, his digital library provides millions around the world with free access to books, music, and movies.

   Three Takeaways:

   - Hobbies can spark innovative ideas.
   - Find joy in what you do.
   - Recreation can lead to impactful projects.

6. **Community and Causes:** Bryan Stevenson,[6] a social justice activist, dedicated his life to helping those wrongly condemned. Bryan grew up in a poor, racially segregated community in Delaware. Inspired by his grandmother, who was the daughter of enslaved people, he developed a strong sense of justice. After studying law at Harvard, Bryan moved to Alabama to represent those wrongly condemned. His dedication to social justice led to the creation of the Equal Justice Initiative, helping countless individuals receive fair representation.

   Three Takeaways:

   - Commitment to a cause can create significant change.
   - Stand up for social justice.
   - Use your skills to serve the community.

7. **Physical and Mental Well-Being:** Dr. Nadine Burke Harris,[7] a pediatrician, recognized the impact of childhood trauma on long-term health. Her focus on mental well-being led to groundbreaking work in addressing adverse childhood experiences (ACEs). Early in her career as a pediatrician in San Francisco's Bayview-Hunters Point, Nadine noticed a disturbing trend—many of her young patients suffered from trauma, which significantly impacted their health. Her observations led her to research ACEs, linking them to long-term health issues. Her groundbreaking work changed the way we perceive and address childhood trauma.

   Three Takeaways:

   - Physical and mental health are interconnected.
   - Early intervention can prevent long-term health issues.
   - Recognize your purpose and the importance of mental well-being.

The stories of these seven individuals, each from different walks of life, are bound by several common themes:

- Resilience in the face of adversity: Each individual faced significant challenges—societal norms, racial barriers, and resource limitations—but they persevered. This resilience helped them break barriers and make significant contributions in their fields.
- Passion-driven pursuits: Each individual was driven by a deep-seated passion, whether it was for medicine, technology, education, or social justice. This passion fueled their dedication and commitment.
- Lifelong learning: Each person displayed an insatiable thirst for knowledge. They were committed to education and con-

tinuous learning, which played a crucial role in their personal development and achievements.

- Impact on society: Each individual used their skills, knowledge, and passion to make a positive impact on society. They didn't just strive for personal success; they aimed to improve lives and make the world a better place.
- Friendships and partnerships: Strong relationships, like the friendship between Akio Morita and Masaru Ibuka, played a vital role in some of these stories. These relationships provided support, shared vision, and collaborative problem-solving.

When trying to write down your values and build a foundation, consider the following:

- Identify your passions: What drives you? What do you care deeply about? Your passions can guide your goals and give you a sense of purpose.
- Embrace lifelong learning: Commit to continuous growth and development. Seek knowledge and skills that will help you achieve your goals.
- Cultivate resilience: Learn to persevere in the face of challenges and setbacks. Resilience is often the key to breakthroughs and success.
- Build strong relationships: Nurture friendships and partnerships that provide mutual support, shared vision, and collaboration.
- Aim for impact: Strive not just for personal success but also for ways you can make a positive difference in your community or the world.

Reflecting on these themes and lessons can help you clarify your values, set meaningful goals, and build a strong foundation for your life.

Questions to Ask Yourself

Over the course of my life, I've had the privilege of engaging with countless individuals in search of betterment. These individuals yearn for success not just in their careers but in life as a whole. I've found that one of the most powerful tools in this journey is self-reflection. It's in asking ourselves the right questions that we uncover our deepest values, guiding us toward fulfillment and true prosperity.

You will find a series of questions inspired by real-life stories of perseverance, passion, and impactful contributions. These inquiries are designed to provoke thought and introspection, especially for those of you in the tech field. They connect to larger themes that resonate universally—empathy, lifelong learning, global citizenship, health and well-being, and environmental stewardship.

Remember, the quality of your life is determined by the quality of the questions you ask yourself. So, as you delve into these questions, I encourage you to be honest, open, and ready to discover new facets of your character.

Here are some reflective questions you can ask to help discover your own values for life and work. Pay attention to the various intentions behind the questions. Thinking and acting in these multiple dimensions will render richer outcomes for you and those around you.

Empathy and Understanding:

- How do I respond to different perspectives and experiences from my team?
- How can I better understand and appreciate the diverse backgrounds of my colleagues?

Awareness and Advocacy:

- What societal issues am I most passionate about?
- How can I use my skills and platform to advocate for these issues?

Lifelong Learning:

- What new skills or knowledge do I want to acquire in the next year?
- How can I foster a culture of continuous learning within my team or organization?

Global Citizenship:

- How do global events and trends impact my work and industry?
- What can I do in my role to contribute to positive change on a global scale?

Health and Well-Being:

- How do I balance work demands with personal health and well-being?
- How can I promote a culture of wellness within my team or organization?

Environmental Stewardship:

- How does my work impact the environment?
- What steps can I take to make my work more sustainable or contribute to environmental conservation?

Resilience:

- How do I handle setbacks or challenges in my work?
- What can I do to build resilience in myself and my team?

Passion-Driven Pursuits:

- What aspects of my work am I most passionate about?
- How can I align my daily tasks with these passions?

Strong Relationships:

- How do I cultivate strong relationships in my professional network?
- How can I improve collaboration and partnership within my team?

Impact:

- How does my work contribute to the betterment of society?
- What changes can I make in my role to increase this impact?

I hope these questions have served as a mirror, reflecting your values, passions, and aspirations. I trust they've stirred thoughts and sparked ideas, leading you closer to understanding your unique path to success. I bet you have better ones in your mind. Write them down and share them.

Remember, it's not the challenges we face that define us but how we respond to them. How we respond is largely influenced by our values. So, keep these questions close, revisit them often, and let them guide you as you navigate the complex, exhilarating landscape of life.

ACTIVITY: Discover and Document Your Values

This exercise is designed to help you explore and articulate your values—the guiding principles that shape your actions, decisions, and relationships. It's an opportunity to dive deep into what truly matters to you across different areas of your life.

The aim is not just to identify these values but also to understand their origins, significance, measures of success, potential challenges, and nonnegotiability. To make this journey more enriching, you'll also be reaching out to your trusted circle for their perspectives.

This activity is meant to be both introspective and interactive. So, feel free to use words, shapes, pictures, or any other creative means to capture your insights. Let's get started!

Instructions:

- **Identify Your Values**: For each category (family, friendship and social life, career and employment, education and personal development, recreation and fun, community and causes, physical and mental well-being), write down two to three values that guide you in that area.
- **Trace Their Origins**: Reflect on where each value comes from. Is it from a personal experience? A family tradition? An influential person in your life?
- **Understand Their Importance**: Write why each value matters to you. How does it influence your decisions, actions, and relationships?
- **Define Success**: For each value, define what success looks like. How do you know when you're living according to this value?
- **Anticipate Challenges**: Consider why and how someone might challenge each of your values. How would you respond to such challenges?

- **Choose Your Nonnegotiables**: From all the values you've listed, select the top five that you will not compromise on, no matter what.

A recommended timeline for this activity is two weeks. Here's a suggested breakdown:

- Days 1-3: Identify your values
- Days 4-5: Trace their origins
- Days 6-7: Understand their importance
- Days 8-9: Define success for each value
- Days 10-11: Anticipate challenges
- Days 12-13: Choose your nonnegotiables
- Day 14: Review and reflect on your insights

During these two weeks, reach out to two to three people from your trusted circle for a chat. Share with them what you're doing, ask for their perspectives, and incorporate their insights into your reflections.

This journey of self-discovery is not always easy, but it's incredibly rewarding. As you navigate through this exercise, remember to take your time, be honest with yourself, and stay open to new insights. Happy exploring!

Don't forget to have fun with this! Utilize your dedicated notebook or journal while you read this book and hurry up slowly. Draw, doodle, or use colorful markers—make it truly yours. This isn't just an activity; it's a reflection of you. So, let your creativity shine!

Ikigai

Let's journey to the East, specifically to Japan, home of a concept that has fascinated and guided millions—Ikigai.

Ikigai, a Japanese term, is most often translated as "reason for being." It's that sweet spot at the intersection of what you love (your passion), what the world needs (your mission), what you are good at (your vocation), and what you can be paid for (your profession).

It's about finding fulfillment, happiness, and balance in multiple areas of life. When you discover your Ikigai, you uncover a powerful source of motivation, propelling you forward on the path of success and fulfillment.

In essence, Ikigai is not just about making a living; it's about making a life—a life of purpose, passion, and joy. Isn't that what we're all striving for?

I invite you to explore this beautiful concept, discover your own Ikigai, and embark on a journey toward a more fulfilled, balanced, and meaningful life.

Sketching Ikigai: A Tale of Two Brothers

As the sun sank low in the sky, casting long shadows across the room, Jake sat across from his younger brother, Sam. The rebellious middle-schooler had a sullen look on his face, a result of one too many visits to the school principal's office. Jake, having navigated through his own share of challenges in college, knew he had to intervene.

"Sam," Jake began, trying to keep his tone light, "do you remember when I was in college and I used to be stressed out all the time?"

Sam shrugged, his gaze fixed on a random spot on the table. "Yeah, so? You were always fussing about grades and stuff."

Jake nodded. "Right. I was trying to control everything, even things that were beyond my reach. It took me a while to understand that it's more important to focus on what's within our control."

Sam scoffed, "And what's that supposed to mean?"

Jake leaned forward. "Things like your actions, your effort, and your attitude. These are all within your control, Sam."

Sam huffed, rolling his eyes. "I don't need a philosophy lesson, Jake."

Jake sighed but didn't let Sam's dismissive attitude deter him. He knew that acceptance of help is often harder than recognizing the need for it. "All right, no philosophy. Let's talk about something else then. What do you love doing?"

"Why does it matter?" Sam muttered, but after a pause, he added, "Drawing, I guess. And I'm not bad at it."

"That's great, Sam!" Jake said enthusiastically. "Now, have you thought about how your love for drawing intersects with what the world needs, or what you could get paid for? Maybe you could make a comic book series or design logos?"

The idea seemed to pique Sam's interest, despite his best efforts to seem indifferent. Jake recognized that glimmer of curiosity. "This, Sam, is the essence of Ikigai, a Japanese concept. It's about finding that sweet spot between what you love, what you're good at, what the world needs, and what you can earn money from."

Sam remained silent for a moment before finally speaking, "So you're saying I should sell my drawings?"

Jake smiled. "If that's what makes you happy, why not? But more importantly, it's about focusing on what's within your control, like using your talent in drawing to do something you enjoy and could potentially benefit others as well."

The room fell into silence as Sam absorbed his brother's words. He was far from accepting everything Jake said, but a seed was sown. Sometimes, that's all it takes.

Through Jake and Sam's story, we see a reflection of our own struggles. We often believe we have all the answers, reluctant to admit when we need help. But by focusing on what is within our control and discovering our unique Ikigai, we can find a path toward a more fulfilled and balanced life.

"Technology is nothing. What's important is that you have faith in people, that they're basically good and smart, and if you give them tools, they'll do wonderful things with them."
—Steve Jobs

Workplace Anecdotes to Understand How Ikigai Helps

In the narratives that follow, we delve into the labyrinth of challenges faced by tech professionals in their quest for fulfillment. We touch upon an elusive search for purpose, where personal values and professional projects oftentimes sit at odds. We then probe into a delicate equilibrium, a tightrope walk between personal passions, societal contributions, professional skills, and financial needs. Next, we venture into the realm of motivation, a vital spark in a landscape that is perpetually evolving and fraught with complex problems. We also navigate through the turbulent waters of high-stress levels and burnout—common companions in an environment of long hours and relentless pressure. Finally, we explore the challenge of maintaining focus in a whirlwind of rapid technological change and incessant multitasking. Each narrative is a piece of a larger puzzle, shedding light on the journey toward finding one's Ikigai in the tech industry. Read along as we unravel these stories, each offering unique insights and perspectives.

Ikigai encourages us to discover and pursue what truly matters to us. When we align our actions with this purpose, we find more joy and satisfaction in what we do. A software engineer at a large tech company was feeling unfulfilled despite a high salary and strong career progression. He realized his true passion was in education technology, so he transitioned to a smaller startup focused on creating educational apps for kids. Even though he took a pay cut, he found his work much more fulfilling because it aligned with his purpose—using technology to educate future generations.

By considering all four components of Ikigai—what you love, what the world needs, what you're good at, and what you can be paid for—we can strive for a balanced life that acknowledges our passions, skills, and societal contributions while also considering our financial needs. A data analyst loved her job and was good at it, but she felt something was missing. She realized that while she loved analyzing data, she also wanted her work to have a greater social impact. So, she started a side project within her company using data analysis to improve the company's sustainability efforts. By aligning her skills and passion with what the world needs and what she could be paid for, she found her Ikigai.

Knowing your Ikigai can provide a powerful motivation to overcome challenges. When we're working toward something we love and believe in, obstacles become less daunting. A team of developers working on a complex project were facing numerous challenges and delays. Remembering their Ikigai—creating innovative software solutions that users love—helped them stay motivated and persevere through the difficulties. Their app eventually became one of the top-rated in its category.

Studies have shown that people who have a clear sense of purpose tend to live longer and healthier lives. This is particularly true in regions known as Blue Zones, where people live significantly longer. One such zone is Okinawa, Japan, where the concept of Ikigai originated. The Chief Technology Officer (CTO) of a thriving tech startup was always busy and under immense stress. Over time, this began to affect his health. He then discovered the concept of Ikigai and realized that while his work was financially rewarding and beneficial to the world, it wasn't something he loved or felt he was particularly good at. He decided to step down from his role and joined a nonprofit, where he could use his tech skills to drive initiatives he was truly passionate about. His stress levels dropped, and his overall well-being improved significantly.

In an age of endless distractions, Ikigai helps us focus by defining what deserves our time and energy. By focusing on what truly matters, we can lead more productive and meaningful lives. A product manager was juggling multiple products and struggling to stay on top of everything. Upon learning about Ikigai, she decided to focus on the one product that aligned with her passion (user-centric design), her skills (product management), what the world needed (the product solved a key customer pain point), and what could pay her bills. By focusing on this one product, she was able to significantly improve its performance and user satisfaction.

In essence, Ikigai is a tool for understanding ourselves better and creating a life that reflects our values and aspirations. By guiding us to find the intersection of our passion, mission, vocation, and profession, it offers a path toward a fulfilling and balanced life.

ACTIVITY: Discover and Document Your Ikigai

STEP 1: Draw four separate boxes on a page in your notebook or journal, each representing one aspect of Ikigai—what you love, what you're good at, what the world needs, and what you can get paid for.

Document your IKIGAI

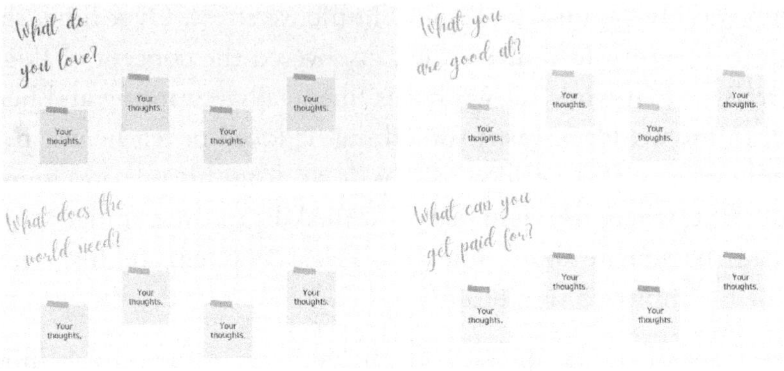

STEP 2: Dive deep into each aspect of Ikigai:

- **Box 1:** What do you love? Think about the activities that genuinely excite you and make you lose track of time. Use words, draw pictures, or even paste photos that symbolize these passions. Examples to inspire you include the ever-evolving field of software engineering. You might love learning new programming languages, frameworks, or technologies. The process of diagnosing and fixing hardware issues might be something you enjoy. You might find satisfaction in setting goals, planning strategies, and seeing them achieved by your team.
- **Box 2:** What are you good at? Reflect on your skills and strengths. These could be technical skills like coding, creative abilities like writing, or soft skills like empathy. Don't limit yourself to what you're currently doing in your job. Remember, this is about discovering your true potential. Examples to inspire you include identifying errors in code and coming up with effective solutions. You might be adept at creating hardware prototypes and iterating on designs based on test-

ing. You might be skilled at communicating clearly and effectively, both verbally and in writing.

- **Box 3:** What does the world need? This can be challenging but think broadly. The world needs sustainable solutions, empathy, education, innovation, and so much more. What are some problems or issues you feel strongly about? Examples to inspire you include technological advancements in healthcare, such as telemedicine platforms, artificial intelligence (AI)-powered diagnostic tools, wearable health monitors, and more to make healthcare more accessible and efficient. Despite advancements, there are still regions around the world with limited access to the Internet. Developing affordable, accessible connectivity solutions can help bridge this digital divide.
- **Box 4:** What can you get paid for? Here, consider the economic aspect. What are skills or services that people or organizations would be willing to pay for? These could align directly with your current job, or they might be tasks you've always considered just a hobby. If you have a passion for biology and computing, you could venture into bioinformatics. This field has a broad range of applications, including healthcare, where technologies are needed to improve diagnostics and treatment. If you're good at building and testing AI models or systems, this could be a lucrative career choice. If you excel at designing and implementing cloud-based solutions, this role could be profitable.

STEP 3: Now, draw four overlapping circles on a new page. Fill these circles with the items from your boxes based on their overlaps. For instance, if there's something you love and are good at, it goes in the overlap between those two circles. Continue this process until all items from the boxes are in the circles.

STEP 4: Look for the items that fall into the overlap of all four circles—this is your Ikigai. These are the things that you love, are good at, the world needs, and you can get paid for.

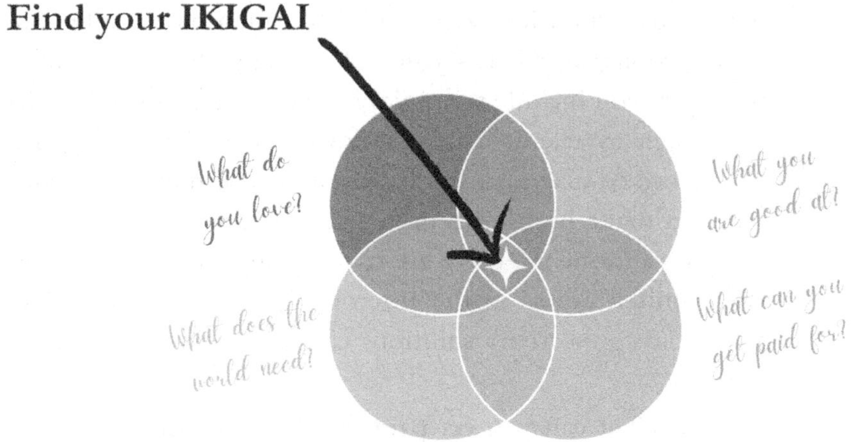

STEP 5: Iteration is the key. You might not find clear answers immediately, and that's okay. Keep revisiting your Ikigai diagram, refining and adding to it as your understanding evolves.

STEP 6: Reach out to a trusted circle of family, friends, and colleagues. Ask them what they think you're good at or what they see as your passions. Their perspectives can provide valuable insights and help you see things about yourself you might have overlooked.

Remember, this is not a one-time exercise. It's an ongoing journey of self-discovery. Be patient with yourself and keep an open mind. There are no right or wrong answers here, only insights that bring you closer to finding your Ikigai.

Finally, set aside a dedicated time each week to work on this exercise. Give yourself at least a month to really delve into it. You could use this as an activity book, adding to it bit by bit, making it a creative and engaging ritual.

By the end of it, you'll have a visual representation of your Ikigai—a guide to leading a fulfilling life that balances your passion, mission, vocation, and profession. Happy discovering!

Section Summary: Purpose

As we journeyed through these pages, we delved deep into the realm of purpose and value and discovered the profound concept of Ikigai. We ventured into the heart of what makes us tick, what propels us forward, and what gives our lives meaning and fulfillment.

We learned that purpose is not just about doing what we love but also about making a difference in the world and leaving a mark that echoes in the corridors of time. It's about finding your true north, that guiding star that lights up your path even on the darkest of nights.

We also explored the concept of value, realizing it's not just about monetary worth or societal prestige. It's about recognizing our innate strengths—our unique gifts—and leveraging them to make a positive impact. It's about understanding that we are each a priceless piece of the cosmic puzzle with an irreplaceable role to play.

Then, we unveiled the treasure trove that is Ikigai—the intersection of what you love, what you're good at, what the world needs, and what you can be paid for. It's our raison d'être, our reason for being, that sweet spot where fulfillment and joy reside. It's not an end goal but a dynamic, ongoing process of self-discovery and growth.

So as we turn the page, remember that your journey toward finding your Ikigai is unique, and it's okay to take your time. Embrace the challenges, savor the victories, learn from the setbacks, and most importantly, enjoy the journey. Because life isn't just about the destination—it's about the beautiful and sometimes winding path that leads us there.

Remember, you are the artist of your life. Each day is a fresh canvas, waiting for you to paint your masterpiece. So pick up that brush and start painting the life you've always dreamed of.

SECTION 2: VALUES-BASED LEADERSHIP

As we embark on this new section, let's kickstart your journey into the world of values-based leadership with an engaging and calming activity. It's a favorite among many students, and I'm sure you're going to find it equally exciting.

In front of you are scrambled words, each representing a key facet of leadership or a value often associated with great leaders. There are few words in here that are associated with negative leadership qualities too. Your mission, should you choose to accept it, is to unscramble these words and identify what they represent.

Are you ready for the challenge? How many of these words can you unravel? Grab a pen and paper, set a timer for 5 to 10 minutes, and see how many you can unscramble. This isn't just a fun game; it's an exercise in reflection, as each word you uncover will shed light on an important aspect of leadership.

aedl	etriindnc	evitaitini	noregv
tnciaap	tnemegaruocne	reegsnit	gaaeennmmt
aohcc	anguceid	lfecniuen	wyas

Once you're done, continue reading, and we'll reveal the answers at the end of this section, keeping the suspense alive. Don't worry if you can't figure out all of them. The process is more important than the result. This exercise is designed to get you thinking about the qualities that define a leader and, hopefully, inspire you to cultivate these in your own journey toward becoming a values-based leader. Let's dive in!

Why Matters: A Revolutionary Leadership Story

In the heart of bustling New York City, there was an esteemed health-care company known as Apex Health Innovations. Apex was globally renowned for its groundbreaking medical advancements and had a reputation that echoed across the industry. Within its modern walls, a transformation was on the horizon.

Enter Rani, an ambitious leader from Australia, still in her early 30s, who was appointed as the new head of Apex's global research and development division. This division was a crucial component of the Apex engine, having been led for over two decades by the esteemed Hiroshi, a man whose medical expertise was renowned throughout the industry.

Rani's appointment was met with raised eyebrows. The team members were all older, more experienced, and had worked under Hiroshi's steady hand for years. Among them were the tenacious Tom, a seasoned medical researcher in his 60s; the spirited Marta, a dedicated lab technician from Italy; and the soft-spoken but astute Chika, a biomedical engineer hailing from Japan. They whispered among themselves, "What does this young woman know about our work? What makes her qualified to lead us?"

Rani, aware of the daunting task before her, recalled the teachings of a mentor she'd never met but who had significantly influenced her leadership ethos—Simon Sinek. She remembered Simon's words, "People don't buy what you do; people buy why you do it."

So, at her first team meeting, Rani didn't talk about changes, improvements, or targets. Instead, she told them why she was there. She shared her belief that the people in a team are more important than the innovations they create, the patients they serve, or the profits they generate. She admitted she didn't have all the answers but believed they could find solutions together. She pledged that their growth and success would be her priority.

Her words resonated in the room, a refreshing change from the usual corporate discourse. The team was surprised and then intrigued. They decided to give Rani a chance.

Under Rani's leadership, Apex's research and development division thrived. More importantly, so did its members. Tom spearheaded a groundbreaking research project that was adopted company-wide. Marta became renowned for her patient-centric initiatives and was promoted to lead a new department. Chika's biomedical solutions revolutionized their processes, earning her cross-organizational recognition.

The real victory lay not in their individual triumphs but in how they became leaders in their own right. Rani became a catalyst, her "why" igniting a spark that transformed the entire team.

This story is a tribute to the power of purpose, of understanding your "why." It's not about your age, your experience, or even your product. It's about why you do what you do. When that "why" is rooted in serving others, in nurturing their growth and success, you create not just a successful team but a legacy of leadership.

Cultivating Safety: The Unspoken Leadership Duty for Corporate Success

A safe environment in a corporate setting is more than just physical safety. It's an atmosphere of trust, respect, open communication, and psychological safety where employees feel comfortable to express their ideas without fear of punishment or ridicule. This environment encourages risk-taking, fosters innovation, and promotes collaboration.

The importance of a safe environment cannot be overstated. When employees feel safe, they're more likely to take risks, share ideas, and collaborate effectively. They're less likely to engage in counterproductive work behaviors like withholding information or avoiding responsibility. This sense of safety also contributes to employee well-being, reduces stress, and improves overall job satisfaction.

Consider the story of a renowned tech company that experienced a significant turnaround. The new CEO didn't focus on sales targets or product innovation. Instead, he focused on creating a culture of trust and safety. As a result, employees began to take risks, share innovative ideas, and collaborate more effectively. In time, these changes led to improved product quality, increased sales, and a significant rise in the company's stock price.

Cultivating a safe environment requires conscious effort from leaders. It begins with leading by example. Leaders must demonstrate honesty, communicate openly, and treat all team members with respect. They should encourage open dialogue, listen actively, and respond constructively to feedback. Providing clear expectations and consistent feedback also helps create a sense of safety.

However, building a safe environment isn't without challenges. Resistance to change, deeply ingrained habits, or prevailing cultural norms can all pose hurdles. Leaders may also face pressure to deliver short-term results at the expense of long-term culture building. To overcome these challenges, leaders must be persistent, patient, and committed to their vision.

Sustaining a safe environment requires ongoing effort. Leaders must consistently reinforce the importance of safety, recognize and reward behaviors that contribute to it, and promptly address any behaviors that undermine it. Regularly revisiting and adjusting policies and practices to maintain safety can also be beneficial.

In the same vein, reinventing the safe environment periodically can keep it from becoming stagnant. This doesn't mean discarding the principles of safety but rather finding new ways to reinforce them. This could involve introducing new team-building activities, revising communication practices, or implementing new feedback mechanisms.

Even Those Not in Leadership Have Impact

In the vast realm of technology, it's not uncommon to encounter challenges with leadership, especially when you're not in a position of authority. These situations can feel daunting, but remember, every challenge presents an opportunity for growth.

First, understand that the concept of a safe environment isn't exclusive to leaders. Every individual in an organization contributes to the overall culture. So, while you might not be in a position to make sweeping changes, you can still influence your immediate surroundings.

But how do you navigate these waters without sacrificing your own well-being? It starts with understanding that you have a voice. You can use this voice to communicate effectively, share your ideas, and express any concerns. Remember, communication is a two-way street. While it's important to speak up, it's equally crucial to listen and understand others' perspectives.

Next, consider the notion of managing up. This doesn't mean manipulating your superiors but rather understanding their goals, pressures, and working styles and aligning your work accordingly. How can you make their job easier? How can your skills contribute to their goals? By understanding these aspects, you can position yourself as a valuable team member while also making strides toward creating a safer environment.

Facing resentment toward management can be hard. It's a natural feeling when expectations are not met or when you feel undervalued. But what if you could turn these feelings into fuel for positive change? Ask yourself—what specific actions or behaviors are causing these feelings? Is there a way you could constructively communicate these issues?

Finally, remember that creating a safe environment is a collective effort. It's about fostering a culture of respect, honesty, and open communication. Even if you're not in a leadership role, your actions can contribute to this culture. By demonstrating these values in your interactions, you can influence those around you and gradually impact the larger work environment.

Navigating the corporate landscape can be challenging, but remember, you have more influence than you might think. By understanding these concepts and applying them in your interactions, you can contribute to a safer work environment, manage up effectively, and turn challenges into opportunities for growth.

Making Others Feel Safe

Making others feel safe, especially in a professional setting, involves creating an environment of trust, respect, and open communication. Here are some practical ways to do that:

- **Empathy**: Understand and acknowledge others' feelings and perspectives. Show genuine interest in their ideas and concerns. Think about a time when a friend went through a breakup. You might not have been experiencing it, but you could understand their pain and offer comfort. That's empathy—understanding and sharing the feelings of others. Lack of empathy can lead to misunderstandings, hurt feelings, and strained relationships. It can create an environment where people feel unheard and unappreciated.

- **Open Communication**: Encourage honest and transparent dialogue. Make it clear that all ideas and opinions are valued and that constructive criticism is welcome. Imagine being part of a project at work. Effective communication is like a team huddle where everyone shares their progress, challenges, and ideas openly to ensure the project's success. Without it, there can be confusion, mistakes, and reduced efficiency. Poor communication can lead to conflict and decreased team cohesion.
- **Consistency**: Be reliable and consistent in your actions and decisions. This predictability can help others feel more secure. It's like going to your favorite coffee shop and expecting your coffee to taste the same every time. In the workplace, consistency means maintaining the same standards and behaviors, which helps in building trust. Inconsistency can erode trust and create uncertainty. People may feel insecure if they don't know what to expect.
- **Respect**: Treat everyone with respect, regardless of their role or position. Avoid favoritism and ensure everyone feels equally valued. Consider a group discussion where everyone has a chance to voice their opinion without interruption. That's respect—acknowledging and valuing each individual's input. Disrespect can cause harm to individuals' self-esteem and create a toxic environment. It can lead to increased conflict and decreased productivity.
- **Trust**: Trust others with responsibilities and show faith in their abilities. This can boost their confidence and make them feel safe to take risks and express their ideas. Just as you trust a friend to return a borrowed book, in a professional setting, you trust colleagues to fulfill their responsibilities and uphold their commitments. A lack of trust can stifle collaboration

and innovation. It can make people feel defensive and cautious, hampering progress and growth.

- **Active Listening**: Give others your full attention when they speak. This shows you value their input and helps build a rapport. Imagine a conversation with a close friend where you're fully engaged and responsive. That's active listening. It shows the speaker that you value their words and encourages open dialogue. Without active listening, important details can be missed, leading to errors and inefficiency. It can also make people feel undervalued and ignored.
- **Positive Reinforcement**: Recognize and appreciate the good work of others. Positive reinforcement can boost morale and create a supportive environment. Remember the feeling when someone praised you for a job well done? That's positive reinforcement. It boosts morale and encourages individuals to repeat good behavior. Without it, people may lose motivation and productivity may decrease. A lack of recognition can lead to decreased morale and job satisfaction.
- **Confidentiality**: If someone confides in you, ensure you keep their trust by maintaining confidentiality. Picture being entrusted with a friend's secret. You wouldn't share it with others, right? The same applies in a professional context. Confidentiality ensures that sensitive information is kept secure. Breaching confidentiality can lead to loss of trust, potential legal issues, and damage to your reputation.

Remember, making others feel safe isn't about grand gestures. It's often the small, everyday actions that have the biggest impact.

ACTIVITY: Journey to Safe Spaces – A Personal Growth Activity Guide

Welcome to an exciting journey of personal growth and learning! These activities are designed to help you develop key skills that are essential in creating a safe and welcoming environment for others. While we encourage you to try all eight activities, we understand that everyone has different time commitments and interests. Therefore, even if you choose to engage with only three or four of these activities, you can still make significant strides in your ability to empathize, communicate openly, maintain consistency, respect others, build trust, listen actively, provide positive reinforcement, and maintain confidentiality. Remember, the goal isn't just to complete these activities but to truly immerse yourself in them and learn from your experiences. So, choose the ones that resonate with you the most and begin your adventure into becoming a better friend, colleague, and leader today!

Empathy (7 days): Choose a day to start. Each day, imagine a different scenario where someone might be feeling a certain way (happy, sad, frustrated, etc.). Draw a comic strip showing how you would react and show empathy. By the end of the week, you should have seven comic strips that show your empathetic responses.

Open Communication (3 days): On Day 1, plan a conversation with a friend or family member. On Day 2, have the conversation, ensuring you express your thoughts clearly and listen actively. After the conversation, write down key points and reactions. On Day 3, review your notes and reflect on the conversation and how open communication made a difference.

Consistency (7 days): Start on a Monday. Choose a task like making your bed or preparing breakfast and perform it in the same manner every day for a week. Keep a journal by your side and jot down your process each day. At the end of the week, reflect on how maintaining consistency felt.

Respect (1 day): Choose a day where you'll encounter many opinions (like during a meeting or family dinner). Make a conscious effort to respect everyone's views. Write down these instances and their outcomes. Reflect on them at the end of the day.

Trust (fourteen days): On Day 1, lend a book to a friend. Draw a picture of you handing over the book. Trust them to return it. On Day 14, when they return the book, draw another picture of you receiving it back. Reflect on how it felt to trust someone with something you value.

Active Listening (7 days): Each day for a week, choose one conversation and practice active listening. Write a short summary of what was said and how it made you feel. At the end of the week, reflect on your experiences and how your relationships benefited from active listening.

Positive Reinforcement (7 days): Starting from Monday, compliment or praise at least five people each day for a week. Use a notebook to capture their reactions using words, doodles, or any other creative means. Reflect on these reactions at the end of each day.

Confidentiality (7 days): On Day 1, tell a friend that you're going to keep a secret they share with you for a week. At the end of the week, draw a sealed envelope to represent that you kept the secret. Reflect on how this practice influenced your relationship.

By following these timelines and narratives, you'll gain hands-on experience with each of these eight important elements and understand their value in making others feel safe and valued. Remember that each day presents a new opportunity to grow and learn. The journey toward creating safe spaces for others begins with you. It's about stepping out of your comfort zone, embracing empathy, and fostering an environment of trust and respect.

Embrace these activities not as tasks but as stepping stones on your path to personal growth. Don't rush the process. It's about progress, not perfection. Remember, the highest reward for a person's toil is not what they get for it but what they become by it. So, let these activities transform you into a beacon of safety and understanding for those around you.

You're capable of more than you know. Believe in your ability to inspire change and watch as you turn into the person you aspire to be. Challenge yourself, learn, grow, and, most importantly, enjoy the journey. After all, success is nothing more than a few simple disciplines, practiced every day. You got this!

"Safety is a basic human need. People with a sense of security and belonging are stabilized for learning, creating, innovating. A group of wonderful people become an exceptional group of people when they feel safe."
—Tim Cook

Holbrook Echoes: The Resounding "No Excuse, Sir"

In the year 2014, my wife and I found ourselves drawn toward a place of stark contrasts. Instead of the well-heeled schools in the Bay Area, we chose to volunteer at Holbrook Junior High, a place where the county was engaged in a relentless pursuit to extricate children from the clutches of meth labs and into the nurturing embrace of classrooms.

Many of these young minds, trapped in middle school, were just discovering the magic of reading and writing, their home environments having denied them these simple pleasures earlier. To aid their learning journey, I collected school supplies, science kits, and various other educational resources from my then-workplace, Xilinx, and sent them ahead to the school.

Upon our arrival, my wife and I were entrusted with two classrooms and an opportunity to address all 200 eighth graders in the school gym. We planned to introduce them to the world of Scratch programming, a novelty straight out of MIT at the time. For the larger assembly, I prepared a speech on leadership, intending to share the stories of transformative figures like Dr. Martin Luther King and Gandhi. My wife planned to talk about computer science and coding, a realm equally accessible to both boys and girls.

Standing before the sea of young faces, I began my speech, but it quickly became clear that my words were falling flat. The restless chatter of the students drowned out my voice. I felt like I was standing in the eye of a storm, and I realized I needed to change course.

I halted my speech, deciding to discard my carefully prepared slides. Instead, I chose to share a story—the story of Bob McDonald's early days at West Point. I narrated how he was treated as an inferior being by his upperclassmen. I told them about the day when a jeep splashed muddy water on his freshly polished shoes, and an upperclassman reprimanded him for his dirty shoes. McDonald knew he had four responses to choose from, but three of them led to severe consequences. It was then that he discovered the most powerful response, "No excuse, sir."

I emphasized to the young minds before me how this phrase embodies the spirit of taking responsibility for one's actions and focusing on what is within one's control. I asked them to echo it back: "No excuse, sir!" The gymnasium reverberated with their voices, their enthusiasm palpable. That moment still gives me goosebumps.

After my wife's equally inspiring speech, we were assigned a student to guide us to our next destination. However, we were intercepted by another teacher who took over the role of our guide. He led us to a seemingly quiet classroom at the far end of the school grounds.

What awaited us there was a sight to behold—a dusty table piled high with books that had clearly been untouched for ages. The stern-faced teacher cleared a spot for us and insisted we sit on the table. When we hesitated, he insisted again, this time more firmly.

Turning to his students, he announced that only people of certain esteemed professions and volunteers like us earned the privilege to sit on that table. His words took us by surprise, leaving us humbled and honored. Then, with a gentle smile, he asked us to share a few words of encouragement with his class.

That day, I thought I was there to inspire these young minds with stories of great leadership. Little did I know I would be the one inspired by the values and leadership exemplified by a teacher in Holbrook, Arizona. It's a lesson in humility and service that continues to resonate with me, a beacon of hope not just for our country but for the world.

"The greatest leader is not necessarily the one who does the greatest things. He is the one that gets the people to do the greatest things."
—Ronald Reagan

Leadership Pillars and Activities to Build Your Repertoire

Leadership is not just about actions; it's about the journey of self-discovery and continuous improvement. As you embark on this journey, it's crucial to understand that leadership starts with you. The six attributes we will focus on are self-reflection, humility, balance, authenticity, diversity and inclusion, and continuous improvement. Don't worry if you feel overwhelmed or unsure. Remember, it's not about being perfect in all areas but becoming extraordinary in what you enjoy and adding value to your career and others.

Working on all attributes at once could make the journey longer and less enjoyable. Instead, seek feedback from your trusted circle to prioritize what to focus on. You're encouraged to work in chunks of 25 minutes, taking breaks in between. You might find yourself in a flow state where time flies by. That's okay. Just ensure a distraction-free environment that allows you to delve deep into this journey.

Self-Reflection

The first attribute we'll explore is self-reflection. It's easy to judge others, criticizing their actions while overlooking our own. However, the flaws and strengths we see in others are a reflection of our own. Before pointing fingers, let's introspect and assess our actions and behaviors. Are we unknowingly projecting our negativity and limitations onto others?

On the flip side, don't forget to give yourself credit. Everyone has someone they admire, be it business tycoons, celebrities, or leaders at work. No one is born great. It takes practice and dedication. What you observe and admire can become part of your repertoire. Surround yourself with people who are good at what you aspire to be and learn from them.

ACTIVITY: Self-Reflection

Let's put the concept of self-reflection into action. Begin by listing the top three values you live by. These are your nonnegotiables that define who you are today. Think deeply about their origins and how they shape you. Do these values resonate with your family, friends, or colleagues? Or are they unique to your individual identity?

Next, list some productive habits you follow. These are habits that make your life meaningful and require effort, time, consistency, and discipline. If you're considering adding new productive habits to your routine, identify three or more habits that will bring you joy and assist in your career development. Remember, there's no right or wrong answer here. You're the final decision-maker.

Now, think courageously and list two or more negative habits that do not serve your best interests. This is where your trusted circle plays a crucial role. They can help identify your blind spots and prioritize which negative habits need to be eliminated. It won't be easy, but remember, it's a marathon, not a sprint. Commit to reducing and eventually eliminating these habits over time.

Remember, we're focusing on the "what" and "why." We'll visit the "how" and "when" at a later stage. Keep this exercise handy as you continue your leadership journey.

Humility

As we delve into various leadership styles, we come across a common thread—humility. This is not a cliché; it's a fact. Great leaders know that a groundbreaking idea can sprout from anywhere and anyone. They foster an environment where everyone feels empowered to contribute to decision-making processes at all levels.

Consider the journey of a first-time manager. Before stepping into this role, they were responsible for their individual tasks, working toward the unified business goal. However, upon becoming a leader, their responsibility shifted from executing tasks to motivating and enabling team members to take ownership of their work. The urge to dive in and do the job themselves might be strong, but a great leader knows the importance of placing trust in their team. This requires humility.

By allowing team members to collaborate, make informed decisions, and execute tasks, teams often achieve better results in the long run. As the philosopher Aristotle taught us, "The sum is greater than its parts." The collective effort of a team often yields a result greater than what could have been achieved individually.

You've probably heard the saying, "There's no 'I' in team." It's a reminder that teamwork is not about individual accomplishments but collective success.

ACTIVITY: Humility

Humility may seem like an abstract concept, but it's simple and transformative once embraced. Globally respected leader Nelson Mandela exemplified humility, teaching us that when you make others feel unthreatened, they're more likely to embrace you.

Reflect on an instance where you experienced or observed humility. It could be an interaction with a colleague, a leader's decision, or even a personal experience. Think about how that instance made an impact. Did it lead to a positive outcome? Did it change your perception of the person or situation?

Next, ponder over the relationship between humility and competition in your environment. Are they seen as opposing attributes, or can they coexist? Humility does not mean being passive; it involves recognizing the value others bring and fostering an environment where everyone's contributions are valued.

Lastly, commit to incorporating more humility in your leadership style. This could involve seeking feedback more often, acknowledging the efforts of your team more openly, or taking a step back to let others shine. List at least one specific goal you will commit to in your journey toward becoming a humbler leader. Remember, leadership is a journey, not a destination. It's about continuous growth and improvement.

Balance

When we think of leadership, we often picture a high-wire act of balancing multiple responsibilities simultaneously. This is not just a skill, but an art—the art of balance. Leaders need to strike the perfect equilibrium in their approach toward handling tasks, teams, and even their personal demeanor.

Leadership is not always about leading. Sometimes, it's about following and trusting your team members to take responsibility. Delegation is crucial, but it can be detrimental if it leans too much or too little. It's a delicate balance that needs to be maintained for the overall success of the team and organization.

In the same vein, leaders are often seen as tenacious individuals who persevere against all odds. While resilience is admirable, there is also strength in asking for help. In today's fast-paced world, where everyone is juggling multiple roles, asking for help is not a sign of weakness but rather an acknowledgment of interdependence and collaboration.

The traditional view of leadership often portrays leaders as the sole decision-makers. While there are situations where this is necessary, in the long-term, it's more effective to decentralize decision-making. A leader cannot be everywhere at once, but a team aligned with the same goals and objectives can distribute executive decision-making responsibilities and back each other. This not only empowers the team but also promotes a sense of ownership and accountability.

Innovation and standardization are two sides of the same coin. While pushing the boundaries of technology and business through innovation is necessary, balancing it with standardization of repetitive processes and tasks is equally important. It's the responsibility of both the leadership and individual contributors to strike this balance.

A key aspect of balance in leadership comes from understanding the difference between assertiveness and aggressiveness. To illustrate this, let's consider a situation where you're standing in line at a coffee shop, and someone cuts right in front of you. A passive response is to avoid confrontation, keeping your anger within you. An aggressive response is to yell at the person, disregarding their feelings completely.

Being assertive, however, is a state of balance. It means respecting your own feelings while also considering the other person's feelings. It's about communicating your point respectfully without devaluing others' perspectives. This powerful technique can defuse difficult situations and lead to positive outcomes for all parties involved.

ACTIVITY: Balance

To understand where you stand in terms of balance, consider a spectrum from passive (with a score of zero) to aggressive (a score of 10), with assertiveness right in the middle (a score of five). Mark yourself on this scale. If you are on either end of the spectrum, you are encouraged to take small steps to move toward a score of five, symbolizing assertiveness.

Next, ask 10 different people to give you their score for you on the same spectrum. Compare this feedback with your self-assessment to see if there are discrepancies between how you perceive yourself and how others perceive you. Owning this perception is crucial, even if it is hard to control. Instead of worrying about it, focus on your own development and learn the skills to reflect the perception you want others to have toward you.

Lastly, reflect on how comfortable and open you are to different viewpoints. For an individual contributor, this means considering inputs from various sources and making your own decision in the end. For a leader, it involves decentralizing decision-making and encouraging diverse perspectives. This openness and willingness to embrace different viewpoints is another crucial aspect of balanced leadership.

Authenticity

Authenticity is often hailed as the holy grail of leadership. The concept, however, can seem elusive and overrated. Instead of striving for authenticity, what if we centered our focus on simply being human? Let's explore this with a real-life example.

Consider a software application engineering manager who was about to meet with her engineering team to discuss product improvements. She knew from past experience that the team would likely attribute user dissatisfaction to poor documentation. This argument previously led to unproductive discussions and hampered collaboration. Determined to pivot from this reality, she started the meeting by acknowledging the issue of ineffective documentation. She expressed understanding and commitment to address their concerns but also requested to table the conversation about documentation for the time being. Instead, she invited suggestions for other product enhancements that could be prioritized in the upcoming release cycles. Her approach was not just authentic but also deeply human, showing respect for both teams' efforts and creating a collaborative path forward.

ACTIVITY: Authenticity

Now, let's delve into what authenticity means to you. Consider a few characteristics that help you define authenticity. These could include honesty, vulnerability, consistency, or any other trait that resonates with you. For each characteristic, assign a level of importance. Some might hold high weight, some lower, and others may fall somewhere in between. There's no right or wrong here; it's about understanding your perception of authenticity.

Next, identify three individuals who embody authenticity for you. They could be leaders, role models, friends, celebrities, or anyone else. Reflect on what makes these people authentic in your eyes. What traits do they exhibit that you admire?

Now, on a scale of one to 10 (10 being the most authentic), assign a score to yourself. This is for your eyes only and serves as a personal benchmark. Regardless of your score, list three specific goals that you will commit to in order to enhance your authenticity. Remember, it's not about achieving perfection but striving for continuous improvement. Being authentic, or rather, being human, is an ongoing journey, a process of embracing our strengths and weaknesses and continuously evolving.

Diversity and Inclusion

In the modern workplace, diversity and inclusion transformed from being buzzwords to integral elements of organizational culture. However, it's essential to understand that diversity goes beyond skin color, cultural background, or gender. It also includes age, experience, and situational diversity.

A millennial-dominated workforce can greatly benefit from the wisdom and diverse life experiences of older-generation employees. Or, a woman who juggles multiple roles—mother, wife, and Chief Financial Officer (CFO) of her family budget—brings a wealth of situational experiences and perspectives. Likewise, hiring a talented individual who lived and worked in multiple countries enriches your team with cultural experience, varied skills, and knowledge.

However, diversity alone is not enough. Inclusion is equally important. Being friendly and welcoming is a good start, but inclusion goes beyond that. It involves creating an environment of mutual respect, rewarding and recognizing diverse contributions, celebrating different cultures, and acknowledging special days off that reflect heritage. It's about implementing detailed strategies that focus on leveraging and valuing differences and what those differences can bring to the organization.

Creating a diverse and inclusive organization must originate in the culture defined and managed by the executive team. It's a feeling that must permeate the entire company. Leaders must focus on the value that individuals with unique perspectives, skills, and knowledge can deliver as a competitive advantage.

ACTIVITY: Diversity and Inclusion

This activity is inspired by the story "The Crayon Box that Talked"[8] and aims to help you understand the concept of diversity and why it should be celebrated.

First, imagine a box of crayons where each color represents a different character with unique traits and perspectives. Here are the descriptions for some of the colors:

- Red Crayon: Loves nature and seeks harmony.
- Green Crayon: Values trust and perfection.
- Blue Crayon: Prefers simplicity and avoids conflict.
- Yellow Crayon: Is always cheerful and optimistic.
- Purple Crayon: Enjoys being unique and values creativity.
- Orange Crayon: Is outgoing, ambitious, and not afraid to stand out.

Initially, these crayons don't get along due to their differences. However, they eventually realize that working together creates something far more beautiful than what they could create individually. This realization teaches them the value of diversity and collaboration, leading them to decide to work together going forward.

Now, your task is to write an essay based on this scenario. Your essay should include the following:

- **Introduction**: Describe the initial state of the crayon box. How do the crayons interact with each other? Why don't they get along?
- **Body**: Discuss the journey of the crayons toward understanding the value of diversity and collaboration. What events lead to their realization? How do their interactions change?
- **Conclusion**: Summarize the transformation of the crayons and the lessons they learn. How does their experience illustrate the importance of diversity and inclusion?

Remember, the aim of this activity is not just to narrate a story but to explore the underlying themes of diversity, collaboration, and inclusion. Use the unique qualities and perspectives of the crayons to illustrate these themes. Good luck!

Continuous Improvement

The path to growth and success is often paved with continuous learning, hard work, and the courage to step outside your comfort zone. While natural talent can give you a head start, it's the relentless pursuit of improvement that separates the truly successful from the rest. Remember, there's always room for growth, no matter where you are in your journey.

Toxic people and environments can hinder your progress. It's crucial to recognize these situations and seek help when necessary. While toxicity should be avoided, discomfort should be embraced. Discomfort signals growth and learning. By becoming comfortable with what makes you uncomfortable, you become a formidable force in your field and a beacon of support for those who depend on you.

Books can become your greatest allies in your journey toward continuous improvement. If you're not a habitual reader, it's never too late to start. The key lies in consistent effort. For instance, reading just two pages a day can gradually transform into a lifelong habit of learning.

The journey to form new habits or learn new skills is not always linear. Consistency, rather than intensity, often yields better results. Research suggests that practicing a new activity three to four days a week is more effective in forming a habit than practicing seven days a week. Remember, the goal is to push yourself without breaking yourself.

Here are some examples of mindset shifts that can aid your journey:

- Old Me: I don't read.
- New Me: I read at least two pages a day.
- Old Me: I cannot learn a new language.
- New Me: I learn Spanish on an app.
- Old Me: I perform tasks manually.
- New Me: I automate routine tasks.
- Old Me: I do not stand up for myself.
- New Me: I'm assertive.
- Old Me: I don't hear other people's opinions.
- New Me: I actively listen.

Incorporating standard templates for recurring tasks, such as to-do lists or data entry, can free up time for tasks that require more focus and creativity.

For further reading on building effective habits, I recommend *Atomic Habits* by James Clear. The book offers practical strategies on how to form good habits, break bad ones, and master the tiny behaviors that lead to remarkable results.

Consider the story of Will Smith, who, along with his brother, was tasked by their father to rebuild a brick wall. Instead of being overwhelmed by the magnitude of the task, they focused on laying each brick as perfectly as possible. Over time, their consistent effort resulted in a sturdy wall. This story is a powerful metaphor for life. Instead of focusing on the enormity of your goals, concentrate on taking small, consistent steps toward them.

Your journey toward continuous improvement is personal. Your competition is not with others but with yourself. Are you better today than you were yesterday? Last week? Last year?

ACTIVITY: Continuous Improvement

For this activity, choose three skills that you commit to developing over the next year. The goal is not to achieve perfection immediately but rather to make consistent progress. Here's a step-by-step guide.

Skill Selection (Timeline: Day 1): Identify three skills that you want to improve or learn. These can be professional skills (e.g., learning a new programming language), personal skills (e.g., improving your public speaking abilities), or even lifestyle-related skills (e.g., cooking, gardening).

Planning (Timeline: Days 2-3): For each skill, define what you want to achieve, how you plan to achieve it, and when you hope to see results. For example, if one of your chosen skills is to learn Spanish, your plan might look like this:

- What: Achieve conversational proficiency in Spanish.
- How: Use a language learning app like Duolingo, practice speaking with a native speaker once a week, and watch Spanish movies or TV shows.
- When: Aim to hold a basic conversation in Spanish within six months.

Execution (Timeline: Day 4 onward): Start working on your skills according to the plans you made. Consistency is key here. It's better to practice a little every day than to practice a lot at once and then burn out.

Milestones (Timeline: monthly check-ins): Set monthly milestones to track your progress. Using the Spanish example, a one-month milestone could be completing the beginner level on Duolingo.

Adjustments (Timeline: as needed): If you're not making the progress you hoped for, don't be afraid to adjust your plans. Maybe you need more practice, or maybe your goals were too ambitious. That's okay. The important thing is to keep going.

Outcome (Timeline: 1 year from Day 1): By the end of one year, you should have made significant progress in your chosen skills. Remember, the goal is not to become perfect but to improve.

Remember to celebrate your progress along the way, no matter how small it might seem. Every step forward is a victory worth celebrating. Good luck on your journey to continuous improvement!

Section Summary: Values-Based Leadership

Answers to the unscramble exercise at the start of this section:

Lead	Direction	Initiative	Govern
Captain	Encouragement	Steering	Management
Coach	Guidance	Influence	Sway

As we conclude our exploration of values-based leadership, a few key takeaways emerge. These principles serve as a compass to guide you on your leadership journey, helping you navigate challenges and make decisions that are in line with your core values:

- **Lead by Example**: Your actions speak louder than words. Demonstrate the values you wish to see in your team.
- **Empower Others**: Great leaders inspire their teams to reach their full potential. Provide them with the tools and support they need to excel.
- **Be Authentic**: Authenticity fosters trust and respect. Be true to who you are and what you stand for.
- **Communicate Effectively**: Clear, open communication is vital. Listen actively and speak honestly.
- **Prioritize Ethical Decision-Making**: When faced with tough choices, let your values guide you. Choose the path that aligns with your principles.
- **Embrace Continuous Learning**: Never stop growing and improving. Seek feedback and use it to become a better leader.
- **Foster a Positive Culture**: Cultivate an environment where everyone feels valued, respected, and motivated to do their best.
- **Serve Others**: Leadership is about serving others, not wielding power. Strive to make a positive impact on your team and your community.

Print this list out and keep it nearby. Look at it daily, perhaps first in the morning or last at night. Let it serve as a reminder of the leader you aspire to be.

Remember, leadership is not about being perfect; it's about being purposeful. It's about making a difference, one decision, one action, one day at a time.

As you embark on your journey of values-based leadership, I encourage you to imagine the kind of leader you wish you had. Then strive, every day, to become that leader. In the end, the best way to predict the future is to create it. So go forth and create a future where you are the leader you always wanted to be. Success is not something you pursue; it's something you attract by the person you become.

"The challenge of leadership is to be strong, but not rude; be kind, but not weak; be bold, but not a bully; be thoughtful, but not lazy; be humble, but not timid; be proud, but not arrogant; have humor, but without folly."
—Jim Rohn

SECTION 3: KNOWLEDGE IS POWER

ACTIVITY: Fruit or Vegetable?

As we embark on the last section of this chapter, "Knowledge is Power," let's dive into a fun and educational activity. Let's challenge our assumptions and expand our understanding with a game called "Fruit or Vegetable?"

The premise of this activity is simple yet enlightening. It revolves around the fact that many produce items we commonly refer to as vegetables are scientifically classified as fruits. For instance, did you know that a tomato is technically a fruit, not a vegetable? Knowledge is recognizing this fact, but wisdom is knowing not to put a tomato in a fruit salad!

Instructions:

- Gather Your Items: Make a list of common produce items—cucumber, eggplant, bell peppers, olives, avocados, pumpkins, etc. Try to include items that you think are vegetables.
- Make Your Guesses: For each item on your list, write down whether you think it's a fruit or a vegetable.
- Research: Now comes the fun part! Use reliable online resources to find out the scientific classification of each item. You might be surprised by what you discover!
- Score Yourself: Give yourself a point for each correct guess. How did you do?
- Reflect: Consider what this exercise tells you about the nature of knowledge and assumptions. Think about other areas in your life where you might be operating under misconceptions.

You can complete this activity at your own pace. However, to truly absorb its lessons, I recommend spreading it out over a week. Spend a few minutes each day researching a couple of items from your list.

This activity is designed to go beyond the pages of this book. It encourages you to seek knowledge actively, question your assumptions, and apply your newfound understanding in practical ways (like making a delicious, scientifically accurate fruit salad!)

Remember, the goal of this activity isn't just to learn about fruits and vegetables. It's to cultivate a mindset of curiosity and lifelong learning. After all, the more we know, the better equipped we are to navigate the world around us. Knowledge and wisdom are two sides of the same coin. Knowledge is about understanding the facts, like knowing that a tomato is a fruit. Wisdom, on the other hand, is about applying that knowledge in a practical, beneficial way—like not putting tomato in a fruit salad.

In life and work, you'll encounter situations where things seem obvious. However, it's crucial to pause, evaluate, and apply wisdom appropriately. Don't rush to conclusions or actions. Take a moment to reflect and consider the best course of action.

As we continue our journey through this section, let this activity remind you that learning can be fun, surprising, and deeply rewarding.

The Real Power of Knowledge

Prepare for a revelation that might surprise you—knowledge is not power. Yes, you read that right. It's a bold statement, and it may go against what you've always believed. You might argue that power is power, but let's delve deeper.

The real power lies in applying knowledge and influencing people. It's not just about knowing; it's about doing and making an impact.

We spend a significant portion of our lives in pursuit of knowledge, from our early years in school to our careers. We learn a myriad of subjects—language, math, science, history, geography, finance, philosophy, and more. We're taught to experiment, use tools, think critically—the list goes on.

However, the amount of knowledge we accumulate isn't the endgame. What truly matters is how we apply this knowledge. How do we translate theory into practice in complex environments? How do we persuade others to leverage our solutions while simultaneously identifying and addressing our own blind spots?

In this section of the course, we will explore various ways of learning. We'll examine traditional methods we grew up with and uncover new strategies that we may have overlooked. Our exploration will be structured and categorized for clear understanding and easy application in real-life scenarios.

So, are you ready to uncover the true power of knowledge? Let's embark on this enlightening journey together.

Seek, Multiply, and Share Knowledge

Learning is a treasure that will follow its owner everywhere. It's often said that those who ask questions risk appearing foolish for a moment, but those who never ask will remain foolish forever. Learning isn't just about accumulating knowledge. It's about questioning what you think you know. The problem isn't that people know too little but that they believe so many things that simply aren't true.

Knowledge has no value unless it's put into practice. Knowing isn't enough; we must apply. Wishing isn't enough; we must do. One of the most effective ways to learn is by acknowledging and learning from our failures. Our journey through knowledge starts with our senses, evolves into understanding, and culminates in reason. Nothing surpasses reason.

Sharing your knowledge is a pathway to immortality. It's not only beneficial to others, but it also reinforces your own understanding. Our interaction with knowledge can be categorized into three main areas:

- Seeking Knowledge: This is akin to self-help.
- Multiplying Knowledge: This involves experimentation.
- Sharing Knowledge: This results in helping others.

As you seek, multiply, and share knowledge, I encourage you to incorporate the following seven strategies.

Strategies to Gain Knowledge

1. Constant Learning: Stay ahead of the curve, no matter the speed or volume of information.
2. Learn How to Learn: Understand the best methods for you to absorb new information.
3. Challenge Your Understanding: Don't be complacent with your current knowledge. Always strive for more.
4. Teach Others: It's a great way to validate and reinforce your wisdom.
5. Ask Questions: Even if you think you already know the answer, you might discover something new.
6. Listen: Do it to gain clarity, not just to respond.
7. Have Fun: If you're not enjoying the process of learning, you're doing it wrong.

Remember, when you seek knowledge, you help yourself. When you multiply knowledge through experimentation, you expand your understanding. And when you share what you learned, you aid others. Some of these strategies apply to one or two areas, while others are universally beneficial. Regardless, they all contribute to making you a stronger, more knowledgeable individual.

STRATEGY 1: Constant Learning

How frequently and for how long must a person exercise and maintain a balanced diet to see results? Say you hit the gym for the first two days of the week but don't notice any immediate change. You might be tempted to think it's not working. But what would happen if you were consistent for six to eight months?

Similarly, how long does it take for a person's health to deteriorate when regular diet and exercise stops? Imagine someone who has been working out for two to three years suddenly stops due to circumstances. Do they experience any immediate change? What would happen if they don't resume their regular diet and exercise within six to eight months?

Spend some time pondering these questions and jotting down your answers.

Next, compare these scenarios with gaining knowledge. How does the example of diet and exercise relate to the process of learning? Write down your thoughts before moving on.

Now, consider how the same analogy applies to nurturing relationships. Again, take your time to write down your thoughts.

Practice makes perfect. Take painting for example. The more you do it, the better you get. The active approach to learning is valuable, especially in tech. For instance:

- Practice coding on platforms like LeetCode or Codecademy.
- Create a personal project, like building a website or an app.
- Participate in a hackathon to turn ideas into reality.

To understand complex human interactions, introspect and assemble experiences like LEGO blocks. Reflecting on experiences can provide profound insights. Consider these steps:

- After completing a project, take time to review what worked well and what could be improved.
- Regularly take a step back from your work to see the big picture and understand how your role fits into it.
- Maintain a journal of your daily learnings and challenges.

Engage in conversations, bounce ideas off each other, and learn from others. Learning from others can open up new perspectives. Here's how:

- Attend tech meetups and conferences to network with professionals.
- Participate in online forums and communities like Stack Overflow or GitHub.
- Collaborate with colleagues on projects to learn different approaches and solutions.

An essential form of knowledge gathering that can't be overstated. Knowledge gained through reading is invaluable. You might want to:

- Follow tech blogs or subscribe to newsletters like TechCrunch or Wired.
- Read books on technology trends, programming languages, or software development methodologies.
- Stay updated with research papers published in your field of interest.

Challenge what you already know. This derivative of other forms of learning is crucial for growth. Questioning what you know can lead to deeper understanding. Try these:

- Engage in code reviews and actively seek feedback.
- Discuss new technologies or methodologies with peers to understand their pros and cons.
- Participate in a debate or a discussion forum on tech topics.

Step out of your comfort zone while respecting your boundaries. It's a superpower waiting to be unleashed. Overcoming obstacles can lead to growth. Some strategies are:

- Learn a new programming language or framework.

- Take on a challenging project or task at work.
- Participate in coding competitions or challenges.

Master the challenges within your mind. It leans heavily toward mental resilience. Conquering your fears can boost your confidence. Consider these:

- Present your ideas or projects to a group or at a meetup.
- Ask questions in a public forum, overcoming the fear of appearing uninformed.
- Apply for a role or a project that you think is slightly out of your reach.

Make substantial statistical data tangible and interesting. Practical application of knowledge can make learning fun and interesting. Here's how:

- Experiment with new tools or technologies in your personal projects.
- Use A/B testing to determine the best user experience for a product.
- Try different problem-solving approaches to find the most efficient solution.

Especially important for leaders or those aspiring to lead, observing people and situations can harness information and aim with accuracy. Learning through observation can be powerful. You might want to:

- Observe how experienced colleagues handle tasks and solve problems.
- Follow industry leaders on social media to understand their thought process and approach.

- Attend webinars and online workshops to see experts in action.

Constant learning involves persistent effort and time. Just like exercising and maintaining a balanced diet, it requires consistency to see results. Whether it's through practicing, introspection, interaction, reading, reasoning, facing challenges or fears, experimentation, or observation, the journey of constant learning never ends.

ACTIVITY: Navigating Methods of Learning

This activity is designed to help you choose and implement two to three learning methods over a period of six months. Follow the steps below to customize this plan according to your needs and preferences.

STEP 1: Choose Your Methods From the nine methods of learning presented, choose two to three that resonate with you the most. For example, you might pick "Practicing," "Reading," and "Facing Challenges."

STEP 2: Set Clear Objectives For each method chosen, set a clear objective. For instance:

- Practicing: Improve coding skills by building a personal project.
- Reading: Stay updated with the latest technology trends and advancements.
- Facing Challenges: Learn a new programming language or framework.

<u>STEP 3:</u> Define Activities Next, define specific activities for each method. For example:

- Practicing: Develop a website or an app using a technology stack of your choice.
- Reading: Subscribe to tech newsletters, or read a book on a new technology every month.
- Facing Challenges: Choose a new language or framework and complete an online course on it.

<u>STEP 4:</u> Timeline Create a timeline for your activities. Here's a sample for the above activities:

Month	**Activity**
1-2	Begin development of personal project (Practicing). Start reading the first book (Reading). Enroll in an online course for the new language/framework (Facing Challenges).
3-4	Continue project development. Read the second book. Apply learned concepts from the course to your project.
5-6	Finalize and review your project. Read the third book. Complete the online course and receive certification.

<u>STEP 5:</u> Track Progress Maintain a weekly log to track your progress. This will help you stay motivated and identify any potential roadblocks.

STEP 6: Review and Reflect At the end of the program, review your progress against your objectives. Reflect on what you learned, what you could do differently, and how these methods contributed to your personal and professional growth.

Remember, learning is a journey. Embrace the process, and enjoy the discoveries along the way!

"I am always doing things I can't do—
that's how I get to do them."
—Pablo Picasso

STRATEGY 2: Learn How to Learn

Learning how to learn is a superpower that can change your life. It's not just about retaining information but understanding how to organize and apply it. Let's dive into the strategies and techniques that can enhance your learning process.

1. Expecting to Teach

The art of conquering knowledge lies not only in learning but also in teaching. The anticipation of imparting what you learned can significantly enhance your understanding and memory recall. This concept, known as "the protégé effect," shifts your mindset toward the material, making you more attentive to the key points and better at organizing the information.

There are several reasons why this strategy boosts comprehension and memory recall. Firstly, the expectation to teach promotes deep processing of information. You delve beyond mere facts, striving to grasp the concepts and their interconnections. Secondly, effective teaching requires clear and logical organization of information.

This helps structure the knowledge in your mind, facilitating future recall. Thirdly, the preparation for teaching naturally leads to the identification of key points and crucial details, reinforcing them in your memory.

2. Writing Over Typing

The art of handwriting, often deemed old-fashioned in today's digital age, holds a unique cognitive charm that can boost memory retention. This seemingly simple act engages different cognitive processes than typing due to the distinct mechanical actions and level of cognitive engagement required. This might make you reconsider reaching for your keyboard the next time you take notes.

There are several reasons why writing by hand can lead to better memory retention. Firstly, handwriting is a slower process than typing, which may be an advantage for learning. It demands a deeper focus on the information, fostering better understanding and recall. Secondly, handwriting involves intricate motor skills that activate parts of the brain related to learning and memory. The physical act of writing creates a motor memory, aiding in easier recall later on.

3. Sleep Reinforces Learning

Sleep, often overlooked and undervalued, is a critical player in the game of learning. It's not just about recharging your body; it's about consolidating your memories and paving the way for new information to be learned. Let's dive into how sleep weaves its magic on our minds.

The process starts with memory consolidation. During sleep, your brain is like a diligent librarian, sorting through the day's experiences. It strengthens neural connections that form our memories and integrate newly learned data into these networks. Next, both rapid eye movement (REM) and non-rapid eye movement (NREM) sleep phases play their parts. NREM sleep, specifically slow-wave sleep, is crucial for factual, episodic, and spatial memory. In contrast, REM sleep helps consolidate procedural and emotional memories.

4. Vary Your Practice Sessions

In the pursuit of learning new skills, it's not just about practicing more but practicing smarter. One method that can accelerate your learning is to diversify your practice sessions. This could mean alternating the types of problems you tackle or varying the skills you work on each session. This strategy, known as "varied practice" or "interleaved practice," may be your secret weapon for mastering new skills swiftly.

So, why does varied practice work? Firstly, it introduces what's known as "contextual interference." This phenomenon occurs when mixing different tasks, which makes learning more challenging, leading to better retention and transfer of skills. Secondly, when you alternate between skills, your brain works harder to recall or relearn the rules associated with each task. This increased cognitive effort can bolster learning and memory.

5. Use Mnemonics

The art of remembering complex information can be made simpler with the aid of mnemonic devices. These tools are like secret codes, helping us link complex information to easy-to-remember constructs. Proven to be effective, mnemonic devices work in harmony with our brain's natural tendencies, making the process of memory storage and retrieval not just easier but also more enjoyable.

So, how exactly do mnemonic devices enhance our memory? Firstly, they create associations between easy-to-remember constructs and complex information. Our brains are naturally adept at making connections, so linking new data to something familiar can boost recall. Additionally, mnemonic devices simplify complex information into manageable chunks or structured formats, making it easier to remember.

6. The Production Effect

The power of memory can be significantly amplified by the act of producing information, such as speaking out loud or writing it down. This practice, often known as the "production effect," can be a game changer in your learning journey. So next time you're revising, consider reading your study materials aloud and witness the positive impact on your memory.

So why does the production effect work? Firstly, when you vocalize or write something, you actively engage with the information. This requires more cognitive effort than merely reading or listening, leading to deeper processing of the information and enhanced memory. Moreover, speaking information out loud makes it more distinctive. Our brains are naturally inclined to remember information that stands out, and by vocalizing it, you're making it unique and, hence, more memorable.

7. Cultivate a Growth Mindset

The power of belief cannot be underestimated, especially when it comes to learning and intelligence. The mindset you cultivate about your ability to learn can significantly influence your trajectory of success. Embracing a growth mindset—the belief that intelligence can be developed—can substantially boost your motivation and achievement.

Our beliefs about our ability to learn can shape our sense of self-efficacy—our belief in our capabilities to succeed in specific situations or accomplish tasks. Higher self-efficacy can lead to greater confidence, increased motivation, perseverance in the face of challenges, and ultimately, greater success. Moreover, if we believe we can learn and improve, we're more likely to be motivated to put in the effort required to achieve our goals. Conversely, if we believe our abilities are fixed and unchangeable, we may give up easily when faced with difficulties. Lastly, those who believe that intelligence is fixed may fear failure and interpret it as a reflection of their inherent abilities. This fear can prevent them from taking on challenging tasks and limit their potential for growth and achievement.

8. Exercise Regularly

The power of regular physical activity extends beyond just physical well-being. It plays a key role in enhancing our emotional state (affect) and mental processes (cognition), leading to a holistic improvement in our quality of life. So, if you're looking to boost your learning capabilities or lift your spirits, incorporating regular exercise into your routine could be the answer.

So, how does physical activity enhance our affect and cognition? Firstly, it stimulates the release of endorphins, chemicals in the brain that act as natural mood elevators. They can help alleviate anxiety and depression and generally improve mood. Secondly, regular physical activity can regulate sleep patterns, positively impacting mood and cognitive functioning. Exercise also increases heart rate, pumping more oxygen to the brain, thereby promoting better brain health and cognitive functioning.

9. Stay Hydrated

The power of water extends far beyond simply quenching your thirst. It plays a critical role in enhancing cognitive performance. It's essential to stay adequately hydrated, especially during your study sessions. So, if you're looking for an easy way to boost your concentration and improve your learning capabilities, remember to drink enough water throughout the day.

So how does staying hydrated improve cognitive performance? Firstly, our brain is approximately 75 percent water, and maintaining an adequate water balance is crucial for its function. Dehydration can alter the brain's structure and function, leading to mood changes and reduced cognitive functioning, including concentration, alertness, and short-term memory. Secondly, water aids in delivering essential nutrients to the brain that are necessary for its functioning. Proper hydration ensures these nutrients are effectively transported.

Just as a jigsaw puzzle is incomplete without every piece, your journey to master the art of learning requires the integration of various strategies. Each strategy is unique and contributes to shaping your learning process. The key is to assimilate these strategies into your routine gradually, observing how each influences your learning process. It's important to remember that the aim isn't merely to accumulate knowledge but to comprehend and effectively apply what you learned.

Start with one strategy. Perhaps you decide to begin by improving your focus. You might try removing distractions from your study space or using techniques like the Pomodoro Technique to enhance your concentration. Monitor how this change influences your ability to absorb information. Then, once you've made focusing a habit, move on to the next strategy. This could be something as simple as taking regular breaks to prevent cognitive fatigue and enhance memory retention.

The beauty of this approach is that it allows you to adapt to new methods gradually, making the transition smoother and more manageable. It also allows you to observe firsthand how each strategy impacts your learning process, providing valuable insights that you can use to further optimize your approach to learning.

The journey to learning how to learn is a process of continuous experimentation and adaptation. Each strategy you incorporate is a piece of the puzzle, bringing you one step closer to becoming an effective learner. Remember, the goal isn't just to learn—it's to understand deeply and apply your knowledge effectively. So, take it one step at a time, observe the changes, and adjust your approach accordingly. After all, learning is not a race but a journey, and every step you take brings you closer to mastering the art of learning.

ACTIVITY: Harnessing Effective Learning Techniques

This activity aims to help you identify your current learning methods, incorporate new techniques, and develop a more effective and efficient learning approach.

Instructions:

- Self-Evaluation: Start by reflecting on your current learning methods. How do you learn today? Do you mostly read, listen to audio materials, or watch videos? Write down your primary learning methods.
- Identifying New Techniques: Next, research and list at least two techniques that can help you learn faster. These could be methods like the Feynman Technique, the Pomodoro Technique, or active recall. Write a brief description of each technique and why you believe it could enhance your learning process.
- Belief Assessment: Reflect on whether you believe you can learn faster and deeper simultaneously. Write down your thoughts and the reasons behind them. Remember, your mindset plays a critical role in your learning journey!
- Commitment Evaluation: Consider if you're ready to demolish the "all-or-nothing" approach. Instead of trying to

learn everything at once, are you willing to learn gradually, integrating new knowledge piece by piece? Write down your commitment.

- Continuous Learning Plan: Lastly, decide how often you'll commit to researching and sharing new ways to learn. Will it be monthly, quarterly, or annually? Make a schedule and stick to it. Continuous improvement is key to mastering the art of learning!

Spend about 15 minutes on each step, totaling approximately an hour to complete this activity. However, feel free to take longer if needed. The goal is thoroughness, not speed.

By the end of this activity, you should have a clearer understanding of your current learning methods, potential techniques to enhance your learning, and a plan for continuous improvement. Use what you've written as a guide and reference point as you embark on your journey to become a more effective learner. Remember, learning is a lifelong endeavor. Stay committed, stay curious, and enjoy the process!

STRATEGY 3: Consistently Challenge Your Current Understanding and Knowledge

This strategy is about embracing change and the evolution of knowledge. It's about understanding that what we know today might be revised or expanded upon tomorrow as we continue to learn and grow. Let's explore this using the example of Pluto's status as a planet.

For many years, Pluto was known as the ninth planet in our solar system. However, this changed in 2006 when the International Astronomical Union (IAU) redefined what it means to be a planet. According to the IAU,[9] a celestial body must meet three criteria to be considered a full-sized planet:

1. It is in orbit around the Sun.
2. It has sufficient mass to assume hydrostatic equilibrium (a nearly round shape).
3. It has "cleared the neighborhood" around its orbit.

Pluto meets the first two criteria but falls short on the third. Despite billions of years in existence, it has not managed to clear other objects of comparable size from its neighboring region in space. As a result, it was reclassified as a "dwarf planet."

Until the mid-20th century, it was believed that the locations of continents and oceans were fixed. Alfred Wegener challenged this with his theory of continental drift in 1912, suggesting that continents were once joined and have since drifted apart.[10] Decades later, evidence from the study of the seafloor led to the acceptance of plate tectonics, the idea that the Earth's crust is made up of plates that move, explaining not just continental drift but also the occurrence of earthquakes and volcanic activity.

This is a common belief that most people heard growing up, but it's actually a piece of World War II propaganda. The British government spread this myth to explain why their pilots had such success in shooting down enemy planes during night raids. They didn't want the Germans to know they had developed radar technology, so they credited the pilots' exceptional night vision to a diet rich in carrots.

Digital labor often goes unnoticed or is underappreciated, especially when it comes to content moderation on social media platforms. Thousands of people work behind the scenes, filtering and removing offensive or harmful content to create a safer online environment. This work can be psychologically taxing and is often poorly remunerated, but it's essential to maintaining the digital spaces we use every day.

"A jack of all trades is a master of none,
but oftentimes better than a master of one."
—William Shakespeare

As we navigate our journey toward knowledge and understanding, it's important to remember that often our preconceived notions or interpretations can be challenged and redefined. This is exemplified in the phrase "jack of all trades," which many of us use without knowing its full context or original meaning.

The term "jack of all trades" is widely used today as a somewhat dismissive label for someone who dabbles in many skills, rather than mastering any one of them. However, this interpretation is a departure from its original connotation.

The phrase was first used to describe none other than William Shakespeare, a man known for his diverse set of skills.[11] He wasn't just a playwright; he also assisted with stage setup, costume design, line memorization, and even directing. The complete phrase is actually "a jack of all trades is a master of none, but oftentimes better than a master of one."[11] Rather than an insult, it was originally intended as a compliment, praising the versatility and adaptability of individuals who possess a wide range of abilities.

The evolution of this phrase's meaning serves as a reminder that specialization isn't the only path to success or self-discovery. Exploring multiple areas, trying new things, and embracing a broad skillset can also lead to valuable insights and experiences.

ACTIVITY: Tech Myths Debunking Challenge

This activity involves identifying, investigating, and debunking common myths or misconceptions within your specific tech field. The goal is to challenge your own preconceived beliefs and, by doing so, discover alternate realities that exist within your professional sphere. The final outcome will be a well-researched presentation to present to your colleagues, friends, and some family members.

STEP 1: Identify the Myth (Weeks 1-2)

Start by identifying a common myth or misconception within your industry. It could be something as simple as a widely held belief about a specific technology, programming language, or method that you always accepted as true but never really questioned.

STEP 2: Plan Your Investigation (Weeks 3-4)

Determine how you will investigate this myth. What sources of information will you use? Who can you consult with? How will you ensure that your investigation is thorough and unbiased? Write out a detailed plan for your research.

STEP 3: Conduct Your Research (Weeks 5-8)

Now, execute your plan. Dive deep into the myth, consult with other professionals, read up on recent studies, and explore alternative viewpoints. Document your findings along the way, making sure to capture different perspectives and counterarguments.

STEP 4: Reflect and Analyze (Weeks 9-10)

Reflect on what you learned. Did your research confirm the myth, or did it debunk it? How does this new understanding challenge your previous beliefs? Write a detailed report on your findings and your personal learning journey.

<u>STEP 5:</u> Prepare Your Pitch (Weeks 11-12)

Prepare a presentation that summarizes your findings and personal growth during this process. Highlight the importance of continuously questioning and learning in your profession.

<u>STEP 6:</u> Share Your Findings (Week 13)

Present your findings and share your experience with your trust circle. This could be a formal presentation at work, an informal talk with friends and family, or even a post on your professional network. Encourage them to question their own preconceived beliefs and explore alternate realities within their fields.

By the end of this activity, you will not only have debunked a common myth (or confirmed it), but you will also experience firsthand the importance of challenging preconceived beliefs and being open to new perspectives. This realization can lead to personal growth and improved problem-solving skills in your professional life.

As we continue our quest for knowledge, it helps us to remain open to challenging our current understanding and beliefs. It encourages us to dig deeper, seek out the full context, and appreciate the nuances that can often be lost in common usage. In doing so, we allow ourselves to grow, learn, and see the world from different perspectives.

<u>STRATEGY 4:</u> Teach Others to Validate and Reinforce Wisdom

In this strategy, we're going to use the professional world of American football as an analogy to demonstrate a crucial concept—aiming beyond your target. Regardless of whether you're an avid fan of the sport, completely unfamiliar with it, or somewhere in between, you'll be able to grasp and apply the lessons from this approach.

Setting the Context:

The first round of the NFL Draft is a significant event, where the top 32 players are selected to join professional teams. These first-round picks receive a minimum annual pay of $10 million, with the top 10 earning at least $20 million, and the top four securing a $30 million annual pay. These rankings and eventual pay grades largely depend on the player's performance at the NFL Combine—a series of physical skill challenges that include throwing accuracy, receiving and route running skills, vertical jump, and the 40-yard sprint.

Aiming Beyond the Target:

Among the many techniques used in training for these challenges, coaches often teach athletes to aim for the 50-yard mark as their finish line during the 40-yard sprint. The idea behind this strategy is to overcome the common tendency to slow down as one approaches the finish line. By mentally extending the finish line, athletes maintain their top speed throughout the race, thereby enhancing their performance.

Application to Learning:

This principle of aiming beyond the target can be applied to learning as well. When you learn a concept through reading, experimenting, and thinking, it's akin to aiming to finish the race at the finish line. However, teaching what you learned to someone else is like aiming beyond the finish line.

The principle of "aiming beyond the target," when applied to learning, especially within the technology sector, can unlock a new dimension of comprehension. This approach challenges professionals to not just absorb information but also disseminate it effectively, thereby deepening their understanding and revealing any hidden knowledge gaps.

For instance, let's consider a tech professional learning a new programming language. Simply reading about it, experimenting with code, and thinking through problems equate to reaching the finish line. However, aiming beyond involves teaching the language's nuances to someone else, which often necessitates simplifying complex concepts and finding effective ways to communicate them. This exercise can unveil subtle intricacies that might have been previously overlooked. Moreover, if the "student" poses a question that the "teacher" can't answer, it uncovers a potential blind spot in the "teacher's" knowledge, prompting further exploration. Consequently, this teaching process transforms passive learning into an active, feedback-driven loop, thereby fostering a more profound understanding of the subject matter.

Consider a project manager in a technology firm who just learned about a new project management methodology, such as Agile or Scrum. They could aim for the "finish line" by simply understanding these methodologies and implementing them in their work. That's a good start, but aiming beyond the target would involve teaching these methodologies to their team.

Teaching the team not only reinforces the project manager's understanding of Agile or Scrum but also opens up opportunities for discussion and questions that may expose any gaps in their understanding. For example, a team member might ask why daily standups are important, prompting the project manager to delve deeper into the purpose and benefits of this practice.

This process could lead to more effective implementation of the methodologies. As team members gain a better understanding from their project manager's teaching, they're likely to apply the principles more effectively in their work, leading to improved team performance.

Some of us may already know this. The question is—are we doing what benefits us and making the time for it?

Activity: Art of Teaching Others

Example topics to choose from:

- Software Development: AngularJS - a software developer could teach a junior member of their team or a peer about AngularJS, focusing on its advantages over other frameworks, its structure, and some basic coding examples.
- Marketing: SEO Techniques - a digital marketer could explain the latest SEO techniques to a colleague from another department or a new team member, using real-world examples of successful SEO strategies and discussing the impacts of these techniques on website traffic and rankings.
- Leadership: Transformational Leadership - a manager could discuss the principles of transformational leadership with their team, providing examples of famous transformational leaders and discussing how these principles can be applied within their own organization.
- Project Management: Agile Methodology - a project manager could teach a new team member about Agile, explaining its key principles, the roles within an Agile team, and how Agile projects are run, using past projects as examples.
- Data Science: Machine-Learning Algorithms - a data scientist could explain different machine-learning algorithms to a non-technical colleague, simplifying complex concepts and providing examples of real-world applications of these algorithms.

STEP 1: Identify a Concept (Timeline: Day 1)

I'll utilize AngularJS as an example, and I encourage you to pick a concept you have recently learned and feel confident about.

STEP 2: Prepare Your Lesson (Timeline: Day 2-3)

Create a lesson plan on how to explain AngularJS to someone else. Include key points like its advantages over other frameworks, its structure, and some basic coding examples. Also, consider ways to make the information engaging and understandable, such as interactive coding exercises or real-world use cases of AngularJS.

STEP 3: Find a Student (Timeline: Day 4)

This could be a junior developer in your team, an intern, or even a non-tech colleague interested in learning about software frameworks. Explain to them that you're practicing teaching as a way to solidify your understanding of AngularJS.

STEP 4: Teach Your Lesson (Timeline: Day 5-6)

Use your plan to teach AngularJS to your "student." Encourage questions and discussion. For instance, they might ask why AngularJS is preferred for certain applications, which gives you an opportunity to delve deeper into its features and benefits.

STEP 5: Reflect on the Experience (Timeline: Day 7)

Consider how it felt to teach AngularJS. Did you find that you understood the framework better? Were there any questions that stumped you, indicating areas where your understanding might be lacking?

Expected outcomes for your reference:

- Software Development: By teaching AngularJS, the software developer could reinforce their understanding of the framework, gain deeper insights into its practical applications, and identify any knowledge gaps.

- Marketing: By teaching SEO techniques, the digital marketer could better understand the impact of these strategies on website traffic and rankings, potentially uncovering new techniques or strategies in the process.
- Leadership: By teaching transformational leadership principles, the manager could reinforce their understanding of these principles and gain new insights into how to apply them within their own team.
- Project Management: By teaching Agile methodology, the project manager could reinforce their understanding of Agile principles, improve their ability to run Agile projects, and potentially discover new ways to apply Agile principles.
- Data Science: By teaching machine-learning algorithms, the data scientist could reinforce their understanding of these algorithms, gain deeper insights into their real-world applications, and identify any knowledge gaps.

The act of teaching not only enhances the project manager's grasp of the new methodology but also facilitates smoother execution of projects, ultimately benefiting the entire team and the organization.

STRATEGY 5: Ask Questions

The power of curiosity manifests by asking questions. It is not just a way of understanding the world—it's a way of changing it. Often, we're quick to respond but slow to understand. Remember that by seeking to understand, you are doing yourself a favor. You understand when you listen, and you understand even more when you ask questions and continue to listen. The idea of asking questions makes perfect sense, but the art of asking effective questions is both challenging and simple.

Here are seven tips for asking effective questions:

- Keep it Simple: Avoid complex or multipart questions that can confuse the listener. During a team meeting, project manager Sarah asked her team, "What do you think of the new software architecture, how might it affect our project timeline, and what are the potential risks?" The team seemed confused. Later, she rephrased her question into three separate ones, making it easier for the team to address each aspect one at a time.
- Don't Predict the Answer: Open-ended questions encourage more thoughtful, comprehensive responses. In a user-testing session, UX designer Tom asked, "You found the navigation easy, didn't you?" This led to participants agreeing without providing much insight. In subsequent sessions, he started asking, "Can you share your experiences with the navigation?" This open-ended question led to more honest, insightful feedback.
- Help People Name What They Already Know: This can spark new insights and deepen understanding. During a brainstorming session, marketing lead Emily asked her team, "What have we learned from our past campaigns that we can apply to this one?" This prompted the team to recall their past experiences and apply that knowledge in creating a successful new campaign.
- Expand Exploration: Encourage exploration of ideas and possibilities beyond the immediate topic. Software engineer Raj was troubleshooting a bug. Instead of just asking his colleague, "What could be causing this error?" he expanded the exploration by asking, "Could there be other parts of the codebase that this bug might be affecting?"
- Investigate Perceptions and Feelings, Not Just the Issue at Hand: This can reveal underlying dynamics and motivations. When team morale was low, Human Resources (HR) manager Lisa didn't just ask, "Why is productivity down?" She

probed deeper by asking, "How are you feeling about your current workload? Are there any issues or concerns that might be affecting your work?"

- Avoid Bringing Attention to Yourself: The focus should be on the person answering, not the one asking. During a client meeting, sales executive Mike refrained from saying, "I think this feature is great. Don't you agree?" Instead, he asked, "What's your opinion on this feature?" This kept the focus on the client and their needs.
- Make Room for Silence: Give people space to think before they answer. Leadership coach Karen was conducting a workshop. After asking, "What leadership style do you think is most effective?" she paused and allowed silence to fill the room. This gave participants time to think deeply before sharing their thoughts.

Imagine walking into your workplace—a space where curiosity is the compass guiding every action. You'd be stepping into a world where questions are the key to unlocking previously hidden insights. Picture yourself in a team meeting, simplifying complex queries to illuminate the path for others. As you conduct user-testing sessions, you're not merely seeking validation of your ideas but inviting a wealth of perspectives through open-ended questions.

In brainstorming sessions, you're not only generating new ideas but also helping colleagues tap into their reservoir of knowledge. While troubleshooting, you're not confined to the immediate problem but exploring the farthest corners of possibility.

Human resources isn't just about managing people. It's about empathizing with them and understanding their emotions and perceptions, not just the tasks they perform. As you converse with clients, you're not pushing your agenda but keenly listening to their needs and their vision. Even in moments of silence, you're creating room for deep thought, for introspection.

STEP 1: Reflect on the Tips (Timeline: Day 1)

Think about these seven tips. Which ones resonate most with you? Which ones do you think could be most beneficial in your conversations?

STEP 2: Commit to Incorporating Questions: (Timeline: Day 2)

Decide how many questions from these tips you will commit to incorporating into your daily conversations. Aim for at least one, but the more, the better.

STEP 3: Practice the Art of Asking Questions (Timeline: Day 3 onward)

During your conversations throughout the day, consciously try to incorporate the questions you committed to. Pay attention to how these questions change the dynamics of your conversations.

STEP 4: Reflect on Your Experiences (Timeline: End of Week)

After a week, reflect on your experiences. Did asking these questions help you to understand more? Did they change your conversations in any way?

By consciously incorporating these questioning techniques into your daily conversations, you can expect a variety of outcomes:

- Deeper Understanding: You might find that you understand people and situations better.

- Improved Conversations: Your conversations might become more insightful and meaningful.
- Increased Empathy: By focusing on understanding others, you might develop greater empathy.
- Personal Growth: As you learn more from your interactions, you might experience personal development.

Remember, the power of asking questions lies not only in the answers you receive but more importantly in the journey of curiosity and discovery you embark upon.

STRATEGY 6: Listen to Gain Clarity, Not to Respond

Imagine an ultimate superpower. Perhaps it's flying, becoming invisible, or remembering everything you read. Now, consider another superpower—active listening. It might not sound as exciting as the others, but its impact can be truly transformative.

We all have people we admire—a family member, a mentor, a friend, an author—for their thoughts, actions, or accomplishments. We value what they've done and how they live their lives. But have you ever considered what you can learn from them? How they make you feel good about yourself? How they make you feel safe?

Active listening is the key to all these experiences. It's about going beyond just hearing words and actually understanding the intent behind them. It's about connecting with others at a deeper level and making them feel valued and heard.

Here's an interesting fact: we speak at 120-150 words per minute, but our brains process at 600-800 words per minute. This means there's room for our minds to wander during conversations. Active listening helps us keep our focus on the speaker and truly understand their perspective.

Suppose you have a friend who is going through a tough time, and they decide to share their feelings with you. As they speak, you put aside your own thoughts and fully concentrate on their words. You maintain eye contact, nod your head in acknowledgment, and refrain from interrupting. You notice their tone of voice and body language to understand the emotions behind their words. When they pause, you ask open-ended questions, like, "How did that make you feel?" or reflect back what they said to ensure you understood correctly, such as "It sounds like you're really stressed because of this situation." By doing so, you're actively listening, showing empathy, and making them feel heard and valued.

Then, imagine you're in a team meeting where your boss is explaining a new project. Instead of thinking about your tasks for the day or checking your emails, you focus entirely on what your boss says. You take notes, ask clarifying questions, and paraphrase their points to confirm your understanding. For example, you might say, "Just to clarify, you'd like us to complete phase one by the end of this month, correct?" After the meeting, you follow up with an email summarizing the key points and action items. This is active listening in action. It not only ensures that you fully understand your tasks but also demonstrates respect for your boss and commitment to your work.

ACTIVITY: The Power of Active Listening

STEP 1: Understand Active Listening (Timeline: Day 1)

Initiate a conversation with a friend or coworker about active listening. Explain the concept, its importance, and techniques such as maintaining eye contact, nodding, paraphrasing, and asking follow-up questions.

STEP 2: Share Personal Experiences (Timeline: Day 2)

Discuss personal experiences related to active listening. Share instances when you felt someone actively listened to you, how it made you feel, and the impact on the conversation. Also, discuss times when you felt someone wasn't actively listening and how that affected the interaction.

STEP 3: Role-Play Scenarios (Timeline: Day 3)

Engage in role-play scenarios where one person speaks while the other practices active listening. Switch roles after each scenario to ensure both of you practice and experience active listening.

STEP 4: Reflect and Discuss (Timeline: Day 4)

Reflect on the role-play exercise. Discuss what you found challenging, what techniques helped you listen more actively, and how it felt being the speaker with someone actively listening.

STEP 5: Commit to Practice (Timeline: Day 5 - Day 10)

For the next week, consciously try to practice active listening in your daily conversations. Pay attention to verbal and nonverbal cues, ask clarifying questions, and paraphrase to confirm understanding.

STEP 6: Reevaluate and Discuss (Timeline: Day 11)

Reconvene with your friend or coworker to discuss your experiences over the past week. What improvements did you notice? What challenges did you face? How did active listening affect your conversations and relationships?

By consciously practicing active listening, you can expect the following potential outcomes.

- Improved Communication: Active listening can lead to clearer, more effective communication, reducing misunderstandings and conflicts.
- Enhanced Relationships: Both personal and professional relationships can improve as active listening fosters empathy and understanding.
- Increased Learning: Active listening can enhance your ability to absorb and retain information, leading to personal and professional growth.
- Better Problem-Solving: With improved understanding of others' perspectives, you might find yourself better equipped to solve problems and make decisions.

Remember, the power of active listening lies not only in the answers you receive but also in the journey of empathy and understanding you embark upon. So, step into that world, fuel your curiosity, and start listening actively. The journey toward personal and professional growth is paved with questions. The more we listen, the more we learn, the more we grow, and the more we empower others. This isn't just a strategy; it's a lifelong commitment to curiosity, learning, and growth.

STRATEGY 7: Have Fun

"If you are not having fun, you are not doing it right." This statement might seem overly simplistic, but it holds profound truth. The journey toward a goal is filled with challenges and requires consistent effort, especially when building new habits. However, finding joy in this journey can be your secret weapon to success.

Unlike happiness, which is often attached to an outcome, joy is embedded within the process. It's the ability to take pleasure in the steps leading up to your goal. When you find joy in performing difficult tasks, it becomes a powerful motivator that propels you closer to achieving your goals.

Let's examine this concept further by exploring the work and corresponding results of different activities, along with their pros and cons:

- Studying: It requires effort and dedication, but it can lead to better grades, a new job, or successful projects. The downside? It can sometimes be tedious and stressful.
- Physical Activities: Working out at the gym, running, or engaging in sports can contribute to a healthier life. However, pushing too hard without proper technique or rest can lead to injuries.
- Socializing: Spending time with loved ones and friends can bring joy, a sense of community, and opportunities for learning and recognition. Yet, if the environment is negative, it could lead to feelings of discredit or stagnation.
- Work Projects: Taking ownership of a project can bring joy and reduce stress when you're passionate about the task. Conversely, factors outside of your control can introduce stress and hamper your joy.

It's clear that every endeavor has its pros and cons. The key lies in focusing on what's within your control and trying to mitigate the impact of external factors. Despite the challenges, it's crucial to nurture your joy and use it as a source of motivation. As you cultivate this joy, you'll find that it not only benefits you but also radiates to those around you.

A great way to amplify joy in your journey is by embracing humor. There are nine different types of humor, and understanding which one resonates with you can help lighten the mood and make the process more enjoyable. Whether it's word-wit, improvisational, or surreal, find what tickles your funny bone and incorporate it into your daily life. Make sure to avoid humor that puts you or others down like self-deprecating, ironic, or dark humor.

ACTIVITY: Unmasking Having Fun

Grab your notebook and write down your thoughts, answers, commitments, and discoveries.

STEP 1: Understand the Concept of Joy and Fun in Challenges (Timeline: Day 1)

Reflect on the statement, "If you are not having fun, you are not doing it right." Consider how joy can be found even in difficult tasks or challenges. Write down your thoughts and experiences related to this.

STEP 2: Analyze Work and Results (Timeline: Day 2)

Identify an instance where you've had to work hard for a result, such as studying for better grades or working out for a healthier life. Note down the pros and cons associated with the results.

STEP 3: Evaluate Your Social Interactions (Timeline: Day 3)

Reflect on your social interactions. How do they contribute to your joy or lack thereof? Are there any changes you could make to enhance the positive aspects and mitigate the negative ones? Write down how negative humor can affect someone. Utilize this to defend quiet ones from the loud ones.

STEP 4: Assess Your Control Over Joy (Timeline: Day 4)

Consider a project or task you've taken ownership of. Did it bring you joy? Were there elements outside of your control that affected your happiness? Write about how you managed or could manage these elements to maintain your joy.

STEP 5: Define Healthy Humor (Timeline: Day 5)

In your opinion, what type of humor is healthy in a conversation? Write down your thoughts and how you believe healthy humor can contribute to a joyous environment.

STEP 6: Embrace Humor (Timeline: Day 6)

Consider healthy humor like word-wit. Identify which types resonate with you and how you respond to those that don't. Reflect on how humor contributes to your joy in your journey.

STEP 7: Evaluate Your Approach to Seriousness (Timeline: Day 7)

Reflect on how seriously you take yourself and others. Is it always, never, or somewhere in between? Explore why you think you hold these attitudes and how they affect your ability to find joy in your activities.

STEP 8: Assertiveness in Humor (Timeline: Day 8)

Consider a scenario where jokes are flying across the room. How would you advise someone to take a joke about themselves but also know how to assertively draw the line?

STEP 9: Complete Your Notes (Timeline: Day 9)

Take your time to think about different scenarios and write them down in your notebook. Reflect on what you learned about yourself and your approach to joy, fun, and humor in challenging situations.

By consciously practicing the art of finding joy in challenges and incorporating humor in your journey, you can expect the following potential outcomes:

- Improved Attitude Toward Challenges: You may find that you approach challenges with a more positive and resilient attitude.
- Enhanced Relationships: Your relationships may improve as you incorporate healthy humor and cultivate a joyous environment.
- Greater Self-Awareness: Through reflection and self-assessment, you will gain a deeper understanding of your attitudes and behaviors.
- Personal Growth: As you learn to find joy in your journey, regardless of the challenges, you might experience significant personal development.
- Better Stress Management: Learning to find joy and humor in difficult situations can be an effective stress-management tool, leading to improved mental health and well-being.

Remember, the journey toward achieving your goals doesn't have to be a solemn, grueling task. Find joy in the process, laugh along the way, and watch how these positive emotions fuel your progress.

CHAPTER SUMMARY

In this first chapter, we embarked on a journey of self-discovery, unearthing our purpose, the importance of values-based leadership, and the infinite power of knowledge. Each lesson was meticulously designed to equip you with the tools needed to become an influential force for innovation, personal growth, and support.

The initial exploration of your "purpose" serves as the compass guiding your voyage. It's this profound understanding of your "why" that will fuel your resolve, particularly during challenging times.

As we progressed, we delved into values-based leadership. Leadership is not merely about authority; it's about inspiring others through your actions. By aligning your actions with your core values, you authentically lead, fostering an environment of trust and respect. This alignment not only propels professional growth but also significantly contributes to personal development.

Lastly, we learned that knowledge is power. In an ever-changing world, the ability to learn, unlearn, and relearn is integral. We must continuously challenge our understanding and embrace the evolution of the mind. This constant pursuit of knowledge keeps us in step with the rapid advances of technology. Moreover, imparting this knowledge to others allows us to aim beyond the target, enhancing our growth and humanity.

Throughout this chapter, we discovered that while doing the right thing is important, finding joy in the process is equally crucial. A simple smile can lighten the journey for you and those around you.

As we conclude Chapter 1, remember that by engaging with these lessons, you've already set yourself apart. You've taken strides toward not only understanding yourself better but also toward influencing the world around you positively.

Your assignment is to encapsulate these insights into a one-to-three-minute video reflecting on what you learned and how this knowledge can enhance your career and life. Use your smartphone and capture your assessment and takeaways—record to teach someone else what you learned.

As we raise our glasses to celebrate this milestone, remember that this journey has just begun. Continue to learn and, most importantly, become the leader you wish you had. Here's to your success and to inspire those who look up to you.

CHAPTER 2

Rising Action - Origin And Current State Of Tension Related To Your Career

This chapter delves into the circumstances surrounding your career, acting as the fertile soil from which your success will sprout. We'll navigate through the inevitable conflicts that arise, viewing them not as stumbling blocks but as opportunities for growth. Finally, we'll confront the challenges that test our resolve and push us beyond our limits. Remember, it's not about where you are now; it's about where you want to go. It's not the conflicts that define you but how you respond to them. And it's not the challenges that hinder you but how you overcome them. As we journey together through this chapter, let's embrace our circumstances, navigate our conflicts, and rise above our challenges, using them as stepping stones toward our success.

SECTION 1: CIRCUMSTANCES

The Paradox of Fortune: A Modern Taoist Tale

An ambitious woman named Li lived in the bustling heart of a modern metropolis. Li was a software engineer at a prestigious tech company known for her dedication and hard work. One day, due to company-wide layoffs, she lost her job. Her colleagues commiserated with her, saying, "Such bad luck!"

But Li shrugged and said, "Maybe."

A few weeks later, while searching for a new job, Li decided to work on an old passion project—an innovative app idea she had shelved due to lack of time. She poured her talent and energy into it. The app caught the attention of a venture capitalist who decided to fund her project. Her friends congratulated her, exclaiming, "What good fortune!"

Yet Li merely smiled and replied, "Maybe."

With the investment, Li's app became a massive success. She became a well-known figure in the tech world, gaining wealth and recognition. When a multinational tech giant offered to buy her app for a huge sum, everyone around her was ecstatic, proclaiming, "Such a great honor!"

As always, Li simply said, "Maybe."

After selling her app, Li found herself entangled in corporate politics and legal battles. The stress and pressure took a toll on her, making her long for the days when she was just an unknown software engineer working passionately on her dream project. She realized then that what seemed like good fortune could sometimes be a mixed blessing and what seemed like misfortune could lead to unexpected opportunities.

Li's journey is a modern interpretation of the ancient Chinese parable that teaches us that life's events are neither intrinsically good nor bad. They just are. It's a lesson that resonates with our course notes—we often label events as good or bad based on immediate outcomes, not realizing that the true impact of these events unfolds over time.

In the intricate dance of life, predicting the consequences of our fortunes or misfortunes is impossible. This understanding should not lead to passivity but rather inspire us to approach life with resilience, adaptability, and a positive outlook.

So, instead of labeling them as good or bad when faced with changing circumstances, remind yourself, "Maybe." Keep moving forward. After all, it's not your circumstances that define you but how you respond to them.

ACTIVITY: The Circumstances Canvas

For this activity, we want to understand the impact of various circumstances on your path to success and to broaden your perspectives through the experiences of others.

Instructions:

- Brainstorm: Start by reflecting on your own life. Identify ten instances that you initially considered good or bad and how they affected your journey toward success. These could be personal, professional, or unexpected events. Examples include losing a job, receiving a promotion, moving cities, etc. Capture these in your notebook assigned for *Hurry Up Slowly*.
- Interview: Reach out to five people from your trusted circle—family, friends, or colleagues. Ask them to share two significant circumstances (one they deemed good and one they

deemed bad at the moment) and how these events impacted their success.

- Capture: Use a large piece of paper or a digital canvas for this exercise. Divide it into two sections—self and others. In each section, further divide it into "good" and "bad." Now, fill in the self section with your own examples and the others section with the examples shared by your trusted circle.
- Visualize: Don't limit yourself to words. Use shapes, pictures, symbols, or any other creative means to represent each circumstance. You could draw a tree for growth, a roadblock for a challenging situation, etc. Let your creativity flow!
- Reflect: Once you fill your circumstances canvas, take a step back. Notice any patterns or common themes. Reflect on how what seemed like good or bad circumstances influenced your journey and that of others.

Day 1-3: Brainstorm and capture your own circumstances.

Day 4-6: Conduct interviews and capture others' circumstances.

Day 7: Visualize and reflect on the circumstances canvas.

At the end of this exercise, you'll have a vibrant, visual circumstances canvas that not only reflects your life's journey but also provides insights into the diverse experiences of others. This is a living document. Feel free to add, modify, or revisit it as you navigate life's circumstances. Remember, in the grand tapestry of life, every circumstance, good or bad, forms an integral thread. Happy mapping!

Life Circumstances

Life circumstances, my friends, are like the various ingredients in a recipe. They mix together to create the unique flavor of your life. Let's explore these key elements.

The environment you work in, whether it's a bustling tech startup or a quiet home office, can greatly influence your productivity and creativity. Picture a coder, who thrives in a silent space, stuck in a noisy open-plan office. It's a recipe for frustration, isn't it?

Geographical influences play a significant role too. Imagine being a software engineer in Silicon Valley where you're surrounded by innovative minds, versus working in a remote town with limited access to cutting-edge technology. The opportunities and challenges you would face are starkly different.

Social media is another powerful factor. It's like a double-edged sword. On one hand, it provides a platform for tech professionals to share ideas and learn from each other. On the flip side, it can also breed a sense of inadequacy when comparing oneself to seemingly more successful peers.

Your socioeconomic status serves as the foundation upon which many other circumstances are built. Consider a self-taught programmer from a low-income background, struggling to afford the latest devices for testing his apps, versus a well-funded tech entrepreneur who has all the resources at his disposal.

Exercise and health are crucial, often-overlooked factors. A healthy body fosters a healthy mind. Picture a young developer experiencing burnout due to neglecting her physical health—her productivity and innovation will inevitably suffer.

Nutrition, too, cannot be ignored. Think about the difference in energy levels between a tech professional who eats balanced, nutritious meals and one who relies on fast food and caffeine to pull through long coding sessions.

Learning and reinforcement are key to staying relevant in the dynamic tech industry. Imagine a seasoned tech professional who resists upskilling versus a newbie who eagerly learns every new programming language—their career trajectories are vastly different.

Recreation, the counterbalance to work, is essential for maintaining mental well-being. Picture two tech gurus—one who spends all his time coding and another who balances his work with hobbies like hiking or painting. Who do you think would be happier and more creative?

Familial influence shapes our attitudes and behaviors. A tech professional raised in a family that values continuous learning approaches challenges differently from someone whose family dismisses the importance of education.

Heredity, too, plays its part. Some people are naturally endowed with a knack for logical thinking and problem-solving—traits beneficial in the tech industry.

Hormones—those tiny chemical messengers in our bodies—can influence our mood and energy levels, affecting our performance at work. A tech professional battling hormonal imbalances might struggle with focus and productivity.

Lastly, gender, in this diverse world, can shape our experiences. A woman in tech, for instance, might face unique challenges in a male-dominated field.

Once you recognize these elements and how they intertwine to shape your life circumstances, you're already one step closer to gaining a deeper understanding of yourself. But awareness is merely the first step. Let's discuss how you can use this newfound knowledge to make actionable changes and see positive results.

If you find that your work environment isn't conducive to your productivity, consider discussing it with your supervisor or HR. You could suggest changes like quiet zones or flexible work hours. For geographical influences, consider virtual networking events or online courses to access opportunities beyond your location.

Use social media to your advantage. Follow thought leaders in your field, join relevant groups, and share your work. However, remember to take regular digital detoxes to avoid comparison fatigue.

If financial constraints hinder your career progression, look for scholarships, grants, or free online resources to upskill. If you're in a privileged position, consider mentoring or supporting less-fortunate peers.

Incorporate regular exercise into your routine. Even a short walk during lunch breaks can boost your mood and energy levels. Remember, a healthy body fosters a healthy mind.

Plan balanced meals and try to avoid excessive caffeine and junk food. Consider meal prepping or subscribing to a healthy meal delivery service if you're pressed for time.

Stay abreast of industry trends and upskill regularly. Participate in workshops, webinars, or online courses. Apply your learning in real-life projects to reinforce your knowledge.

Find a hobby outside of work that brings you joy and relaxation. This will not only reduce stress but also foster creativity and a fresh perspective.

If you come from a supportive family, leverage their guidance. If not, seek mentorship from industry leaders or colleagues. Remember, your chosen family can be as influential as your biological one.

While you can't change your genetic makeup, understanding it can help you play to your strengths and work on your weaknesses. For instance, if logical thinking comes naturally to you, consider roles that leverage this skill.

Listen to your body. If you suspect hormonal imbalances, consult a healthcare professional. Regular checkups, a balanced diet, and stress management can help maintain hormonal balance.

Understand the unique challenges and advantages your gender may present in your field. Seek support from relevant groups and contribute toward making your industry more inclusive.

Change takes time. Start small—pick one or two elements you want to focus on first. As you start seeing positive results, gradually incorporate more changes. Over time, you'll notice a significant improvement in your career and personal growth. It's a journey of self-discovery and self-improvement, so go easy on yourself and celebrate every small victory!

"Your current life circumstances are the results of decisions made and actions taken in the past, whether 10 years ago or 10 minutes ago. They are only circumstances. They do not define you."
—Dean Jackson

So, dear readers, consider these elements as you navigate your tech careers. They intertwine to create your unique life circumstances, influencing your journey to success. But remember, they are just circumstances—they don't define you. You have the power to leverage them, learn from them, and shape your destiny.

ACTIVITY: Empowering Self-Evaluation and Understanding Life Circumstances

This activity will focus on empowering self-evaluation and understanding life circumstances. Being critical of ourselves is natural, but harsh critique can diminish our confidence and hinder personal growth. Instead, we'll focus on acknowledging what needs improvement and celebrating our victories. This activity will help you identify and understand your life circumstances, how they influenced your journey, and how you can leverage them for personal and career growth.

Instructions:

- Identify Areas for Improvement: Identify three or more areas in your life or work that you believe need improvement. The key here is to focus on areas for growth, not on what you've done wrong. Write these down in your notebook.
- Document Life Circumstances: List three to five life circumstances you believe have worked against you. These could be anything from your socioeconomic status, geographical location, health, family dynamics, etc.
- Expand on Challenges: Pick three life circumstances you listed that worked against you and expand on each. Write about why you believe it still exists or how you overcame it.
- Identify Favorable Circumstances: Now, switch gears and list three to five life circumstances that worked in your favor. Again, these could relate to any aspect of your life.
- Expand on Advantages: Choose three favorable life circumstances and expand on each one. Write about why you believe it still benefits you and/or how you can multiply its impact.

This activity is designed to be reflective and thought-provoking, so take your time with it. We recommend setting aside a quiet hour to complete it. However, break it into smaller chunks over a week if that works better for you.

By the end of this activity, you should have a clearer understanding of your life circumstances and how they shaped your journey. You'll also learn to shift your focus from harsh self-critique to empowering self-evaluation. This exercise will help you identify personal and career growth areas and understand how to leverage your favorable circumstances. This is a continuous process so revisit this activity regularly to reflect on your progress and realign your goals.

ACTIVITY: Sharing Your Life Circumstances – Creating Safe Spaces

We invite you to share your life circumstances with us in this activity. The aim is to create a safe space for you to express yourself, especially if you typically don't speak up in gatherings, be it large or small. Remember, everyone's story matters and can inspire others. You never know how your experiences may resonate with someone else, sparking a connection or even helping them in their journey.

Instructions:

- Reflect on Your Life Circumstances: Reflect on your life circumstances that significantly influenced your journey. According to your perception, these could be positive, neutral, or negative.
- Express Your Feelings: For each circumstance, express how it makes you feel. This exercise isn't about judging your feelings but acknowledging them.
- Choose Your Medium: Decide how you'd like to share your story. You can write it down in Microsoft Word, record an audio file, or even create a video.
- Share Your Story: Share with your family and friends. Share your story with us when you're ready. You can email your document, audio file, or video to info@grounow.com. Alternatively, tag us on our social media platforms (www.facebook.com/grounow, www.instagram.com/surajethirajan/, or www.linkedin.com/company/Grounow).

Take your time with this activity. It's important to reflect deeply and express your feelings honestly. There's no rush. Share your story when you feel ready.

By sharing your story, you're creating a safe space for yourself and contributing to a community where everyone's experiences are valued. Your story can inspire others, give them a sense of belonging, or even help them navigate life circumstances. Plus, expressing your feelings can be incredibly liberating and therapeutic. So, step out of your comfort zone and share your story—you're braver than you think!

Professional Circumstances

Life in the workplace is like a game of chess. You have to know the rules, understand the pieces, and always think a few steps ahead. But don't worry. I'm here to help you understand the different pieces and how they move.

You know how, in a video game, you can control your character but not the entire game? That's what work is like. Sometimes, you might wish you could change the game, but focusing on how to play your character better is more effective. For instance, you're a software developer, and the project requirements keep changing. Instead of feeling frustrated, focus on what you can control—your code quality, clarity in communication, and ability to adapt quickly to changes.

Ever tried building a LEGO tower, and it kept falling over? It's frustrating, right? But you didn't give up. You kept trying different ways until it stood up. That's called being flexible and resilient. Imagine you're a data analyst, and your model predictions aren't accurate. Instead of giving up, you try different algorithms, tweak parameters, or gather more data. Remember, there's always a solution if you're flexible enough to find it.

Imagine building a LEGO set without the instructions. Confusing, huh? That's why having clear goals at work is essential. It's like having the instructions for your LEGO set. As a project manager, you must know the end goal and the steps. If instructions are unclear, ask questions until you have a concrete plan. Having clear goals is like having a roadmap to success.

Remember when your favorite video game had an update, and everything seemed different? However, it wasn't so bad once you understood why the changes were made. Suppose your company decides to shift to a new technology stack. Initially, it might seem challenging, but understanding the reasons behind this change can help you embrace it more readily.

Now, imagine if you had to explain the video game update to your friend. Being able to do that at work is essential, too. Imagine you're a team lead and need to explain these tech changes to your team. Doing so effectively will ensure everyone is on the same page.

The rules and mood of the playground (or your workplace) affect how much fun you have. A good mood makes for a happier day at work. As a UX designer, the work culture significantly impacts your creativity. A positive environment fosters creativity, so try contributing to a healthy work ambiance.

Friends at work are like having buddies on the playground. They make work more enjoyable and teamwork easier. As a software engineer, collaborating with your peers can make problem-solving more effective. Building solid relationships with colleagues can make work more enjoyable and productive.

If you and your teacher get along well, you're likely to enjoy class more. The same goes for work. A good relationship with your boss can make work a lot more pleasant. Having a supportive supervisor can significantly enhance your learning experience as an intern. Maintain open communication and respect to foster a positive relationship.

Just like how you must balance school and playtime, adults must balance work and personal life. It's essential to have time for both.

As a tech professional, balancing work and personal life is crucial. Setting boundaries, like not checking work emails after hours, can help maintain this balance.

Imagine you've got a big homework assignment. It might seem overwhelming, but breaking it down into smaller tasks makes it manageable. The same strategy works at work. As a product manager, you may have multiple tasks at hand. Breaking them into smaller tasks can make it more manageable and less overwhelming.

Remember when you were learning to ride a bike? You kept falling, but you knew you could do it. Eventually, you did! That's called self-efficacy—believing you can achieve what you set out to do. You might be tasked with resolving a complex bug as a junior developer. It might seem daunting, but believing in your capabilities can empower you to find a solution.

You know how your character gets stronger every time they overcome a challenge in video games? That's resilience—getting stronger with every problem you solve. As a tech professional, you'll face challenges regularly. Each problem you solve makes you stronger and more experienced. Remember, challenges are growth opportunities.

Life isn't just about reaching the destination. It's about navigating the journey. It's easy to let life happen to you, to be a spectator in your own story. But I'd like to challenge you to take the reins to become the author of your narrative. This exercise might seem simple, but its power lies in its simplicity. It allows you to step back to look at your professional landscape from a bird's-eye view. It allows you to identify the mountains that block your path and the rivers that propel you forward. Don't let skepticism hold you back. I encourage you to dive into the following activity with an open mind and heart. Sometimes, the most remarkable revelations come from the simplest of exercises. I urge you to put aside your doubts, pause whatever you're doing, and embark on this journey of self-discovery. Remember, the only limits in life are the ones we set for ourselves.

ACTIVITY: Navigating Your Professional Circumstances

In the game of life, understanding the playing field is crucial. Your professional circumstances shape your career journey, whether they work for or against you. This activity aims to help you identify these circumstances and explore how to navigate them effectively. It's not just about the destination but also the journey.

Instructions:

- Identifying Challenges: Think about your current professional situation. List three to five circumstances that you believe are working against you. These could be anything from a micromanaging boss to a lack of growth opportunities.
- Overcoming Challenges: Choose three circumstances from your list. For each one, write down why you believe it's a hurdle and brainstorm ways to overcome it. Every challenge is an opportunity in disguise.
- Identifying Opportunities: Now, shift your focus to the positive aspects of your professional life. List three to five circumstances that you believe work in your favor. These could be your supportive colleagues, flexible work hours, or resilience.
- Leveraging Opportunities: Choose three circumstances from your list. For each one, write down why you believe it's beneficial and how you can multiply its impact. Making the most of what you have is crucial to success.
- Sharing Your Insights: Share your insights once you complete the exercise. You can write them down in a document, record an audio file, or make a video. Email your responses to info@grounow.com or tag us on Facebook, Instagram, or LinkedIn. Feel free to share any circumstance—positive, neutral, or negative. Let us know how it made/makes you feel.

We recommend taking at least one week to complete this activity. Spend a couple of days on each step, giving yourself enough time to reflect and introspect.

By the end of this activity, you should better understand your professional circumstances. You'll gain insights into what's working for you, what's working against you, and how you can navigate these circumstances effectively. This exercise will help you understand your professional landscape better and equip you with the tools to navigate it more effectively. Every circumstance is an opportunity to learn, grow, and become a better version of yourself.

Section Summary

You just navigated through Circumstances, the initial part of Chapter 2 - Rising Action.

The power of awareness is transformative. It enables us to redefine and disarm our limitations. Often, what binds us isn't concrete but a tale from yesteryear or a phantom of our imaginings.

Circumstances can alter instantly, yet humans evolve at a more measured pace. Even those who crave change might stumble upon obstacles that make transformation more complex than we initially thought. But don't let that deter you. Persist. Objectively describing your circumstances disconnects you from your constraints. Know this—the character, not circumstances, shapes us.

Cheers to becoming the leader you wish you had and for others to look up to. Keep going; you are doing brilliantly.

SECTION 2: CONFLICTS

True peace doesn't mean avoiding conflict; it means learning how to navigate it in a peaceful manner. Conflict often sparks innovation as it encourages human interaction, argument, and debate. The more adept we become at engaging, listening, interpreting, and responding constructively, the more likely we are to utilize conflict to our advantage rather than being overwhelmed by it.

A skilled leader doesn't seek to eliminate conflict but rather aims to prevent it from depleting the team's energy. If your team openly challenges you when they believe you're wrong, that's a sign of a healthy dynamic. Remember, no pressure, no diamonds.

Effective communication isn't just about hearing the words that are said but also understanding the unspoken. The art of reading between the lines is a lifelong pursuit of wisdom.

The more we evade conflict, the more control it gains over us. The less we fear it, the less it confounds us. The more we acknowledge our differences, the less they divide us.

Now, let's dive right in. This first exercise is designed to help you understand your relationship with conflict. Credit goes to *The Coward's Guide to Conflict* by Tim Ursiny for this insightful activity.

ACTIVITY: Your Conflict Score12

For this exercise, you'll answer 12 questions on a scale of one to five. A score of one indicates that the statement doesn't describe you at all, while a score of five indicates that it's a precise definition of you. The lowest possible score is 12, and the highest is 60.

Instructions:

- Read each question carefully.
- Reflect honestly on how much the statement applies to you.
- Assign a score between one and five for each question.
- Add up your scores to find your total.

Note: This exercise is not about judging yourself but rather gaining insight into your relationship with conflict.

Questions:

1. I often avoid discussions that might lead to conflict.
2. I find it difficult to express my feelings when angry.
3. I struggle to stand up for myself when I feel wronged.
4. I tend to give in during disagreements to maintain peace.
5. I worry about others' opinions of me after a confrontation.
6. I feel uncomfortable when others express anger or disagreement.
7. I often contemplate conflicts long after they've ended.
8. I struggle to say "no" even when I want to.
9. I often feel misunderstood or unheard in conflicts.
10. I have difficulty letting go of resentment after a conflict.
11. I worry that standing up for myself might damage relationships.
12. I find it challenging to stay calm and composed during confrontations.

Once you complete the exercise and calculate your score, reflect on what it reveals about your relationship with conflict. This is not a test but a tool for self-discovery. The aim is to understand yourself better and learn how to navigate conflicts more effectively.

Understanding Your Conflict Score

Once you've added your scores, let's break it down into three simple categories. This will help you understand where you stand and what steps you can take to improve your approach to conflict.

Category 1: 12-24 - The Peacekeeper

If your score falls in this range, you're a peacekeeper. You tend to avoid conflict and value harmony above all else. That's not a bad thing! However, sometimes, standing up for yourself is necessary.

Next Steps: Try expressing your feelings in a respectful way when you're upset. It's okay to disagree sometimes! Practice saying "no" when you need to. Remember, it's okay to prioritize your needs.

Category 2: 25-48 - The Balancer

If your score is in this range, you're a balancer. You understand that conflict is part of life but might struggle to handle it. You might worry about others' opinions or hold on to resentment after arguments.

Next Steps: Practice letting go of past conflicts and focus on the present. Work on listening to others and expressing your views clearly. It's okay to have different opinions, and it's important to respect others' views too.

Category 3: 49-60 - The Challenger

If your score is in this range, you're a challenger. You're not afraid of conflict and are comfortable expressing your thoughts and feelings. However, be careful not to overwhelm others with your assertiveness.

Next Steps: Practice patience and empathy. Listen to others' perspectives and try to understand their feelings before responding. Everyone has a right to their opinion. It's not always about who is right or wrong but about understanding each other better.

Regardless of your category, remember that dealing with conflict is a skill at which we can all improve. It's all about understanding, respect, and communication. So, keep learning, growing, and practicing!

Choosing Your Path in Workplace Conflicts

Conflicts are akin to hurdles in a race. They're a normal part of any professional environment, like meetings or project deadlines. Two team members won't always agree on everything—that's perfectly alright! The secret isn't to fear or sidestep these hurdles but to learn how to cross them constructively.

If we don't tackle conflicts effectively, they can strain our professional relationships. But if we handle them respectfully and positively, they can strengthen our bonds with colleagues. Whether you're experiencing disagreements in the office, during virtual meetings, or in team projects, learning these skills can help you navigate differences healthily and build more fruitful professional relationships.

It's like standing at a junction when you're in a conflict. You have to choose which road to take. Here are some routes you could choose:

- Avoidance: Working remotely to avoid a heated meeting.
- Acquiescence: Accepting a colleague's idea even when you don't fully agree with it.
- Passive Aggression: Expressing your displeasure indirectly, like through sarcastic comments.
- Dominance: Trying to enforce your ideas over others.
- Competition: Vying for the best presentation slot to get noticed by the boss.
- Compromise: Agreeing to adopt parts of your and your colleague's ideas for a project.
- Collaboration: Brainstorming to find a solution that satisfies everyone.

- Respect: Acknowledging your colleague's perspective, even if you disagree.

Now, there's no right or wrong choice here. Each of these routes can be useful at different times. For example, sometimes, it's essential to avoid conflict to maintain a peaceful work environment.

To help you decide which route to take, consider how much you want to achieve what you desire (this is called "engagement") and how much you want to help your colleagues achieve what they want (this is called "cooperation").

Imagine a graph where one axis shows your level of engagement and the other shows your level of cooperation. You can place each conflict choice on this graph to help you decide which route to take.

While it might seem like collaboration, which involves a high degree of engagement and cooperation, is always the best choice, let's explore why other choices might sometimes be more appropriate in real-life workplace situations. After all, professional life isn't always as straightforward as a graph.

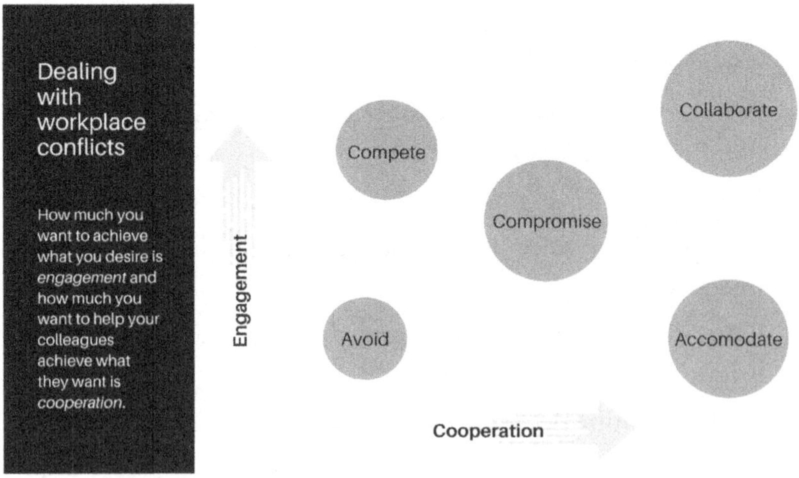

It's important to give credit where it's due. The concepts of "engagement" and "cooperation" in conflict resolution and the ways to approach conflicts mentioned earlier are based on the Thomas-Kilmann Conflict Mode Instrument.[13] This model was developed by psychologists Kenneth Thomas and Ralph Kilmann to help people understand how different conflict-handling styles affect interpersonal and group dynamics and to learn how to choose the appropriate style for a given situation.[13]

Fact Versus Story

Fact and story often intertwine in our minds, blurring the line between objective reality and our subjective interpretation of it. Facts are objective and verifiable pieces of information, while stories are narratives we construct to make sense of these facts. As humans, we inherently crave meaning and are biologically wired for storytelling. We use stories to understand the world around us, predict future events, and connect with others. However, this tendency can sometimes lead us astray when we mistake our stories for facts.

When we confuse stories with facts, we risk misjudging situations and people based on our preconceived narratives rather than objective realities. This can lead to misunderstandings, reinforce biases, and hinder effective communication. For example, in a negotiation, if we formed a story in our mind that the other party is unreasonable or untrustworthy, we might overlook opportunities for compromise and mutual benefit, focusing instead on defending our position.

Furthermore, this confusion can limit our ability to engage with others fully and authentically. If we're wrapped up in our own narrative, we may fail to listen actively, missing out on the nuances of others' experiences and perspectives. This can create barriers to empathy, understanding, and connection, all of which are essential for successful interpersonal relationships and win-win outcomes.

Here are five examples of facts versus stories to illustrate the difference and how they can influence conflicts:

Fact: Your colleague didn't reply to your email.

Story: You might think, "My colleague is ignoring me because they don't respect my ideas." This story can fuel feelings of resentment and conflict.

Fact: A friend arrived late to your dinner party.

Story: You interpret their lateness as a sign that they don't value your time or friendship. This story can lead to unnecessary tension and misunderstanding.

Fact: Your partner forgot to take out the trash.

Story: You might create a narrative that your partner doesn't care about shared responsibilities or your comfort, leading to feelings of anger and dissatisfaction.

Fact: A team member disagrees with your proposal in a meeting.

Story: You may perceive their disagreement as a personal attack and believe they're trying to undermine you, which can spark conflict and hinder collaboration.

Fact: Your child hasn't cleaned their room despite being asked multiple times.

Story: You might interpret this as a sign of defiance or laziness, which could lead to unnecessary friction and punitive measures.

In each of these examples, the fact is a simple, verifiable piece of information. The story, however, is an interpretation or assumption that we layer onto the fact. It reflects our beliefs, fears, and biases more than objective reality. When we confuse these stories with facts, we risk escalating conflicts based on misinterpretations.

So, while stories are powerful tools for making sense of our world, it's crucial to distinguish them from facts. By doing so, we can navigate our interactions with greater clarity, openness, and effectiveness, fostering healthier relationships and more beneficial outcomes for all involved.

ACTIVITY: Deconstructing Conflicts

In the tech industry, conflicts are as common as coding bugs. We often find ourselves embroiled in disagreements and misunderstandings as we navigate our professional lives. If not addressed correctly, these conflicts can disrupt our work environment and impede productivity.

However, it's essential to remember that observing a conflict is more straightforward than navigating through it. By transporting ourselves into an observer's shoes, we can gain a fresh perspective, making it easier to deconstruct the conflict and understand its roots.

This exercise will explore four conflict scenarios you might encounter as a tech professional. This will help you understand how conflicts arise and give you insights into how your reactions might contribute to the situation.

Instructions:

- For each scenario, list the parties involved directly and indirectly.

- Write down each person's perspective. Consider their concerns, frustrations, fears, anger, excitement, happiness, satisfaction, desires, etc. related to the situation.
- Distinguish between fact and story in the situation. Fact is an objective reality (e.g., "The sky is blue"), while story is a subjective interpretation (e.g., "The sky is beautiful").
- Reflect on how your reactions, actions, or inactions might have affected the situation.

Spend about 20 to 30 minutes on each scenario. Spread this across multiple days if necessary.

By the end of this exercise, you should better understand the dynamics of conflicts and how your actions can influence them.

Scenarios for Study:

1. Task Conflict: The Documentation team is frustrated with the Application team and Engineering organization for not providing source content on time for writers to do their jobs and align with the go-live date for the products.
2. Leadership Conflict: A previous manager of over 10 years is replaced by a junior manager. The team was not given any justification from upper management for the change.
3. Creative Idea Conflict: A social media company campaigns to address environmental awareness and preservation. The UI designer spends two weeks on a mock-up. The head of product spends 10 minutes during the review meeting, rejects the idea, and fails to provide constructive feedback.
4. Personality Conflict: Two engineers are assigned to the same marketing campaign. Engineer #1 approaches these tasks with lots of preparation, while Engineer #2 prefers to spend less time preparing but still completes them. This leads to

conflict because Engineer #1 feels that Engineer #2's lack of preparation impacts the quality of their shared work.

Reflect on your past or current experiences and identify two scenarios across the types of work-based conflict. Apply the same approach as above to these personal situations.

This exercise will help you become more self-aware and equip you to handle conflicts more effectively in your professional life. Remember, understanding the dynamics of a conflict is the first step toward resolving it.

Section Summary

In this section, we jumped into the heart of conflicts, peeling back the layers to reveal their true essence. Conflicts are not just about clashing views or interests. They're fertile ground for growth and understanding. Every conflict is a chance to learn and expand your horizons.

We explored how facts and stories intertwine in our minds, often leading us astray. Facts, the indisputable truths of life, and stories, our interpretations of these truths, can often be mistaken for one another. It's here, in this confusion, that the seeds of conflict are sown.

When we mistake our stories for facts, we risk misjudging situations, reinforcing biases, and creating unnecessary division. We close doors on compromise and mutual understanding. The good news is that we can change it once we recognize this. We can navigate our interactions with greater clarity and openness by distinguishing facts from stories.

So, as we journey through life, let's consider separating the wheat from the chaff and the facts from the stories. Let's strive to see things as they are, not merely as we perceive them. With this wisdom, we can transform conflicts from hurdles into opportunities, fostering healthier relationships and more beneficial outcomes for all involved.

Effective communication involves understanding the difference between what is said and what is heard. Ultimately, it's not about winning an argument but understanding each other better. After all, we are not islands but interconnected beings in this vast ocean of life.

SECTION 3: CHALLENGES

Challenges are the spice of life. They are the mountains we must climb and the rivers we must cross on our journey toward our goals. In the realm of work, challenges may appear as roadblocks hindering our path to success. Regardless of our role or level, we all face challenges. Our workplaces—a melting pot of diverse management styles, personality types, and personal objectives—often become the breeding ground for conflicts and challenges. Moreover, an individual battles their unique challenges, perhaps stemming from inexperience or the quest for new opportunities.

You'll encounter numerous challenges throughout your career, much like stepping stones on your professional path. Each step brings fresh expectations and hurdles to surmount. Preparing for these challenges and acquiring strategies to conquer them can lead to success.

Remember, maintaining a positive outlook is crucial. It fuels your motivation and focus, empowering you to tackle any challenge, whether job rejections or conflicts with colleagues. Sometimes, seeking help becomes necessary, be it requesting a friend for a mock interview or asking a supervisor for skill training. Such assistance can bolster your confidence.

Transitioning from a recruit to a seasoned employee conversant with the business's policies, expectations, and culture can be quite a journey. Also, every time you switch companies, you embark on this journey anew. Some common challenges new hires face include conflicts with coworkers, unclear job expectations, and difficulties in completing work tasks.

Promotions, though exciting, bring their own set of challenges. Just like being a new hire, you'll need to adapt to new tasks and responsibilities. Feelings of inadequacy or an overwhelmingly busy schedule might follow a promotion.

If your career trajectory includes managerial roles, you'll encounter unique demands. Managing people responsible for the work can be daunting but incredibly rewarding when you lead others to success. Challenges such as terminating an employee or not meeting performance goals can be mitigated with the right coaching and leadership tools and a professional and committed mindset toward achieving practical business goals.

Life is a roller coaster of uncertainties. You can either embrace it, savoring the ride and learning joyfully from your experiences, or resist life's challenges, resenting every moment of your journey. The former enriches you with growth and development, while the latter robs you of these opportunities.

Challenges are an integral part of life and can be leveraged to your advantage. Each challenge presents an opportunity for personal growth and self-improvement. The ultimate goal is to harness what you learn from these challenges to become the best version of yourself.

"Better" signifies improvement. No matter where you stand today, there's always room for enhancement. Even a monk strives daily for self-improvement, continually aspiring to become a better person.

Your mind shapes your perspective, influencing how you perceive and interact with the world. Conquering the mind is a lifelong endeavor, requiring consistent effort and complete awareness of this infinite task.

Rest assured, you're now equipped to tackle these challenges with grace and strategy. As you continue on your journey toward self-improvement, remember to let go of circumstances beyond your control, stay present in the moment, create space for healthy self-reflection, reassess your values and beliefs, embrace the resilience that setbacks offer, allow failures to foster compassionate resonance, and let loss teach you about what you truly value.

ACTIVITY: Itemize Your Challenges

Welcome to a simple yet profound exercise to help you understand and prioritize your life challenges. This activity is about creating clarity, gaining perspective, and recognizing the power of your mind in shaping your reality. Let's dive into it.

Instructions:

STEP 1: From the following list, identify one to five challenges that resonate with you but are currently missing from your list of hurdles—demanding situations, a weak support system, negativity, failures, overthinking, anxiety, clutter, fear, prioritizing others over us, peer pressure, procrastination, taking things personally, thinking, discipline, overcoming your story, physical abilities, life challenges, mental abilities, toxic people, what others think, lack of knowledge, self-doubt, setbacks, uncertainty.

STEP 2: Gather your top 10 challenges, including those you just identified. Rank them, with one being the most challenging.

STEP 3: Share your ranked list with three or more trusted individuals. Ask them if they agree with your list and have them rank the challenges they agree with.

Allocate a week for this exercise. Spend the first two days on Steps 1 and 2, the next three days on Step 3, and the last two days on analysis and reflection.

After completing this exercise, take some time to assess the responses and results. You may find discrepancies between how you perceive your challenges and how others perceive them. Remember, it's not about whose perception is right or wrong but about gaining insight into different perspectives.

Depending on your personality, you might be tempted to take feedback personally. But other people's opinions are just that—opinions. They need not define your reality.

Use this feedback to identify areas you believe need to change to align with the perception you want others to have of you. More importantly, uncover your blind spots and thoughts holding you back.

Our minds are mighty and can sometimes trick us into believing something unreal. This exercise allows you to identify and prevent these illusions from hindering your progress.

Every challenge is an opportunity for growth. Embrace this process with an open mind and heart and prepare to discover new insights about yourself and your journey.

Understanding Control Through Maslow's Hierarchy of Needs

Our relationship with control is intricately woven into our perception of and response to challenges. The concept of control, in this context, has two pivotal aspects. Firstly, it pertains to our inherent need for it—the sense of stability and predictability that we derive from being in control. Our thoughts, behaviors, and overall mindset often evolve depending on whether we feel in control.

Secondly, control also relates to our psychological propensity to focus more on what we lack, rather than appreciating what we have. This tendency can skew our perspective, causing us to overlook our strengths and opportunities while magnifying our perceived inadequacies and threats.

To better understand this complex interplay of control and challenges, we must delve deeper into our fundamental needs as human beings. For this, Abraham Maslow's hierarchy of needs serves as an illuminating guide.

In 1943, a pioneer of humanist psychology, Maslow proposed a model delineating five primary human needs arranged in a hierarchical order. This model has since been widely recognized and applied across various fields.[14]

A key feature of Maslow's model is the hierarchical nature of needs. The lower the needs in the hierarchy, the more basic and essential they are. As such, individuals tend to prioritize meeting these lower needs over the higher ones. For instance, when unwell, the primary concern becomes recovery, often superseding other concerns about societal perceptions or personal ambitions.

Let's explore Maslow's five needs from the bottom up:

- Physiological Needs: These are the most basic human needs for physical survival and well-being, such as food, water, and health. When these needs are unmet, all other concerns become secondary until we regain our health.
- Safety Needs: Once physiological needs are satisfied, safety needs come into play. These encompass our need for security, stability, and freedom from fear. We seek to control our environment to ensure our safety, using resources like wealth, power, and relationships.
- Belonging Needs: This level of need refers to our desire for interpersonal connections and a sense of belonging. We strive

for acceptance, affection, and affiliation with groups, seeking control over our social standing and relationships.
- Esteem Needs: These needs relate to our desire for respect from others and self-esteem. They involve the need for achievement, competence, recognition, and status. We seek control over how we are perceived and valued by others.
- Self-Actualization Needs: At the top of the hierarchy, these needs represent the desire to realize our full potential and become the best version of ourselves. Control, in this context, pertains to personal growth and self-fulfillment.

Understanding these need layers can help us identify what is within our control. It allows us to focus on tangible aspects we can influence while accepting the elements beyond our control. This balance is crucial in navigating life's challenges and striving toward self-actualization.

Unmasking Blind Spots: The Undeniable Impact of Maslow's Hierarchy in Leadership

Winning a debate does not necessarily mean you are right. It often means you successfully argued your point of view or convinced others of your stance. However, in leadership and management, particularly for those resistant to ideas like Maslow's hierarchy, it's crucial to understand that winning an argument isn't the ultimate goal. The focus should be on achieving a favorable outcome for the team and the organization. For managers and leaders entrenched in traditional, task-oriented approaches, the importance of addressing Maslow's hierarchy of needs may seem irrelevant or unnecessary. However, many hard facts and compelling examples underline the undeniable link between attending to these needs and business success. Here's how we can make this clear to even the most stubborn leaders:

- Quantifying the Impact: Research consistently demonstrates a direct correlation between employee well-being and productivity. For instance, a study by the University of Warwick[15] found that happiness led to a 12 percent spike in productivity, while unhappy workers were 10 percent less productive. Leaders inadvertently hamper productivity by neglecting their team's basic human needs.
- Highlighting Real-World Examples: Companies renowned for their attention to employee needs often outperform their competitors. Google, known for its focus on employee well-being (from physical needs like food and health to self-actualization needs like personal growth), consistently tops lists of the best workplace places and enjoys high employee productivity.
- Addressing the Cost of Neglect: Ignoring employees' needs can lead to high turnover rates, which are expensive for businesses. SHRM (Society for Human Resource Management) reports that replacing an employee can cost up to twice the employee's salary when considering recruiting, training, and lost productivity costs.[16] Leaders can foster loyalty by attending to employees' needs, reducing turnover and its associated costs.
- Emphasizing the Link to Customer Satisfaction: Employees who meet their needs are more likely to provide better customer service. A report from Gallup[17] found that units with engaged employees (those who feel their needs are met) saw 10 percent higher customer ratings and a 20 percent increase in sales.
- Stressing the Importance of Adaptation: Adaptability is key in today's rapidly changing business environment. Leaders who cling to outdated, monolithic approaches risk being left behind. Embracing a more people-centric leadership style can help businesses stay competitive.

- Providing Evidence of Improved Team Dynamics: Teams led by people-oriented leaders tend to have better collaboration and communication, leading to improved problem-solving and innovation. A *Journal of Applied Psychology* study[18] found that supportive leadership practices could significantly improve team problem-solving effectiveness.

Armed with these hard facts and examples, even the most resistant leaders can see the undeniable value of addressing their team members' needs according to Maslow's hierarchy. It's not just a "nice to have." It's a strategic imperative for business success.

"We don't have to be in command at all times. What we truly seek is a feeling of control."
—Suraj Ethirajan

ACTIVITY: Self-Assessment Using Maslow's Hierarchy

This activity will help you evaluate your current state based on Maslow's hierarchy of needs. By identifying areas of strength and those requiring attention, you can gain a clearer perspective on your overall well-being and what steps you may need to take for self-improvement.

Instructions:

- Familiarize Yourself: Start by understanding Maslow's hierarchy of needs, which consists of five levels: physiological, safety, love/belonging, esteem, and self-actualization.
- Assessment: For each of these layers, identify two or more aspects that are going well for you and two or more aspects that demand more attention. Be as specific as possible.

- Reflection: Write down why you chose these specific items. What makes them stand out as either positive or needing improvement?
- Emotional Check-in: Reflect on how these identified items influence your emotions, thoughts, and behavior. How do they make you feel, and in what ways do they shape your actions?

This activity can be carried out over a week. Spend a day or two familiarizing yourself with Maslow's hierarchy and then dedicate a day for each level of the hierarchy for the assessment and reflection process.

By the end of this activity, you should have a comprehensive picture of your current needs and health grade. You will be aware of the areas in your life that are thriving and those that require your attention. This insight can guide your personal growth journey, helping you prioritize improvement areas and nurture your strengths. Remember, self-improvement is a continuous process; this activity is a stepping stone toward better understanding yourself.

Happy self-discovery!

What We Control Versus What We Do Not Control

In life, there exist elements we can control and those that are beyond our reach. The key to success, well-being, and overall joy lies in correctly categorizing these elements and focusing our energy on what we can influence. No matter your level of expertise, let's set the wheels of your mind in motion.

Consider these examples of elements within your control:

- Your Self-Talk: Your inner dialogue holds immense power. If you're prone to self-doubt or negative self-talk, it's time for a change. How you converse with yourself, particularly when others doubt you, is a vital life skill you can master.

- Your Circle: Who you choose to associate with is entirely within your grasp. Surround yourself with winners. As the saying goes, birds of a feather flock together. By aligning yourself with successful individuals, you're more likely to achieve success yourself.
- Your Sleep Routine: You have the power to establish a healthy sleep routine. For instance, avoiding screens an hour before bedtime might be challenging, but it's simple and essential for your well-being.
- Your Readiness to Seek Help: It's within your control to ask for assistance when needed. If you're already doing this, great! Continue to do so and encourage others to understand that it's perfectly okay, even necessary, to seek help.
- Your Diet: Within your means, you control what and when you eat. It requires discipline to resist tasty yet unhealthy options, but we'll explore tools and techniques for building positive habits later on.
- Your Boundaries: Setting boundaries, whether with strangers, friends, or family, is crucial. Remember the airplane safety rule—secure your oxygen mask first before assisting others. Establishing boundaries can be difficult, but we'll learn methods to make it less daunting.
- Your Personal Development Time: The time you spend on self-improvement is under your control. Despite work, personal chores, and other responsibilities, developing discipline and habits to allocate time for self-growth can make you unshakeable.

Consider these seven controllable elements that can significantly impact your life—your thoughts and beliefs, your kindness toward others, your effort, your pursuit of knowledge, your gratitude, how you treat others, and who you choose to surround yourself with.

The way you approach your personal life often mirrors how you approach your professional life. The elements within your control, such as your thoughts, beliefs, and the company you keep, are not limited to your personal world. They extend to your workplace as well.

Take for instance your self-talk. The way you communicate with yourself directly impacts your confidence and performance. If you constantly belittle yourself, it will reflect in your work output and interactions with colleagues. On the other hand, positive self-talk can enhance your problem-solving skills and make you a more effective team player.

Similarly, setting boundaries at work is equally important. This could mean avoiding office politics, setting clear expectations with your manager, or simply saying no when you're overloaded with work. By setting boundaries, you maintain a healthy work-life balance and prevent burnout.

Your readiness to seek help also plays a crucial role in the workplace. It's not a sign of weakness but a testament to your ability to recognize when you need assistance. It leads to increased learning, better problem-solving, and, ultimately, improved performance.

Just as you are mindful about your diet for your physical health, be mindful about your mental diet too. What are you feeding your mind? Are you learning new things? Are you staying updated with industry trends? Your mental diet can greatly influence your growth and success at the workplace.

Lastly, the time you spend on personal development can have a direct impact on your professional growth. The skills you develop, the knowledge you gain, and the habits you form during this time can make you more efficient, productive, and successful at work.

In essence, your life is like a garden. You cannot control the weather, but you can control what you plant, how you nurture it, and when you harvest. Similarly, you can't control all aspects of your job, but you can control your actions, reactions, and attitudes. As you sow, so shall you reap. So, plant good habits, nurture positive thoughts, and harvest success.

ACTIVITY: Discovering What You Control Versus What You Don't

In our journey through life, we often encounter circumstances beyond our control. However, focusing on what we can control can empower us, leading to personal growth and positivity. This activity will guide you in identifying areas of your life that are within your grasp and how you can leverage this control for self-improvement.

Instructions:

1. Thoughts and Beliefs: Identify three or more thoughts and beliefs that you intend to control. These could be things you wish to do more of or less.
2. Kindness Toward Others: Identify two or more additional ways you plan to extend kindness toward others, even if you consider yourself to be already kind.
3. Effort: Identify two or more areas where you intend to increase your effort. These can be personal goals, career objectives, or even hobbies.
4. Pursuit of Knowledge: Identify two or more actions you will take in your pursuit of knowledge. Think about adding to your current practices.
5. Gratitude: Identify three or more people or situations for which you are grateful. Consider sharing a gratitude journal with a loved one, friend, or colleague. Regularly jot down two to three elements or qualities you appreciate about this

person or situation. Aim to do this daily, weekly, or at least monthly.

6. Treatment of Others: Identify two or more actions you will take to treat others better. This goes beyond simple acts of kindness. For instance, you could strive to be a more active listener or to be less judgmental.
7. Company You Keep: Identify two or more actions you will take to refine the company you keep. This could mean spending more or less time with certain individuals, protecting your personal time, or even spending more time with yourself.
8. Reflection: Document your thoughts, responses, and commitments to these actions.

This activity can be carried out over two weeks. Spend the first week on steps 1-4 and the second week on steps 5-8.

By the end of this activity, you'll have a clearer understanding of what aspects of your life you can control. You'll also have a roadmap of actions to improve these areas. This exercise should empower you to take charge of your personal growth journey, helping you focus on what truly matters and letting go of factors beyond your control.

Remember, the power to change your life lies within your hands. Embrace it!

Mental Health

The state of your mind is as crucial as the health of your body. Your emotional, psychological, and social well-being, collectively known as mental health, influences how you think, feel, and act. It shapes how you handle stress, relate to others, and make choices that affect your life.

It's essential to understand that poor mental health and mental illness aren't always the same. A person might struggle with mental health issues without having a diagnosable mental illness. Conversely, someone diagnosed with mental illness can also experience physical, mental, and social wellness periods.

The connection between mental and physical health is profound. Conditions like depression can increase the risk for various physical health problems, including long-lasting ones like diabetes, heart disease, and stroke. In the same vein, chronic physical conditions can heighten the risk of mental illness.

Statistics[19] show that over 50 percent of individuals will be diagnosed with a mental illness or disorder at some point in their lifetime. Every year, one in five Americans experiences a mental illness. Furthermore, one in 25 Americans has a severe mental illness, such as bipolar disorder or major depression.

Mental health can fluctuate over time, influenced by numerous factors. When life's demands exceed a person's resources and coping abilities, their mental health may suffer. For instance, working long hours, caring for a family member, or dealing with financial struggles can all lead to deteriorating mental health.

Online screening is a quick and easy way to determine if you're experiencing symptoms of a mental health condition. Tests cover a range of conditions, including depression, anxiety, psychosis, PTSD, and ADHD, among others. If you're in crisis, contact a trained crisis counselor or call 911 immediately.

Crisis can manifest in many forms, not just thoughts of self-harm. It can encompass any painful emotion or any situation where you need support. Remember, it's okay to reach out for help.

People often struggle to identify their feelings accurately. Most people resort to generic responses like "good" or "bad." However, delving deeper and identifying what we're genuinely experiencing can enhance our communication and relationships.

If negative thoughts or cognitive distortions plague you, exercises are available to combat them. You can also find tips for boosting your mental health, such as maintaining a gratitude journal, working on your strengths, and practicing forgiveness.

In the end, remember this: "You don't have to see the whole staircase. Just take the first step." If you are in crisis, don't hesitate to seek help immediately. Your mental health is not a luxury—it's a necessity.

Print the following for easy and fast access in the case of a mental crisis for you or someone else who needs help.

Call 1-800-273-TALK (8255) to reach a 24-hour crisis center available through the National Suicide Prevention Lifeline.

Text MHA to 741741 to connect with a trained crisis counselor from Crisis Text Line.

Call 911 or go to the nearest emergency room.

If you need support but are not in crisis, consider contacting a warmline. Warmlines offer a place to call when you need to talk to someone. Speaking to someone on these calls is typically free, confidential, and run by people who understand what it's like to struggle with mental health problems.

Find a warmline at mhanational.org/warmlines.

Mental health is a critical part of overall health. If you're feeling distressed, there is hope. Check out https://mhanational.org/ for more details.

ACTIVITY: Exploring and Addressing Mental Health

Mental health is an integral part of our overall well-being. Despite societal progress, there's still a long way to go in acknowledging and addressing mental health issues. This activity aims to help you identify ways to support and develop your own mental health and, if possible, extend that support to others.

Instructions:

- Personal Commitments: Identify three or more commitments you can make to support and develop your own mental health. These could be daily practices, lifestyle changes, or therapeutic activities.
- Supporting Others: Once you establish a strong mental health baseline for yourself, think about how you can extend this support to others. Identify three or more commitments you can make to help others with their mental health. Always ensure you're in the right mental space yourself before extending support to others.
- Recognizing Warning Signs: Identify three or more ways you can help yourself and others notice warning signs of poor or deteriorating mental health. This could be changes in behavior, mood swings, or withdrawal from social activities.
- Combating Negative Thoughts: Identify three or more strategies you can employ to combat negative thoughts. These tactics could include cognitive behavioral techniques, mindfulness exercises, or positive affirmations.
- Seek Support: Reach out to family, friends, colleagues, or your coach to review your responses regarding mental health. Their feedback can provide valuable insights and suggestions to improve your mental health strategy.

This activity can be done over two weeks. Spend the first week focusing on steps 1-2 and the second week on steps 3-5.

By the end of this activity, you should have a clear action plan to support your mental health and, if applicable, assist others. This proactive approach will not only help you manage stress and maintain a positive outlook but also enable you to extend empathetic support to others.

Remember, caring for your mental health is not a sign of weakness. It's the first step toward a happier and healthier life. Don't hesitate to seek help if you need it. You're not alone in this journey.

Confronting Challenges: Letting Go, Seeking Perfection, and Unique Challenges

The road to success is paved with challenges. Learning to navigate these hurdles requires courage, self-awareness, and resilience. In this section, we'll explore three key aspects: letting go, the pursuit of perfection, and addressing unique challenges that may not be covered in this book.

Letting Go

Often, our greatest obstacle is not an external force but a self-imposed limitation. We cling to past failures, old habits, or outdated beliefs, hindering our growth. The act of letting go can be liberating, setting us free to pursue what truly brings us joy.

Consider a software developer stuck on an old coding methodology. By refusing to let go and adapt to new techniques, they limit their efficiency and innovation. However, once they release their grasp on the old method, they open up mental capacity and time to master newer, more efficient techniques, accelerating their professional growth.

Being Perfect

Perfectionism often masks a fear of criticism or disappointment. Many of us would rather disappoint ourselves than others, suppressing our feelings and desires to maintain harmony. This tendency is particularly prevalent in the tech industry, where the pressure to deliver flawless results can be immense.

For instance, a project manager striving for perfection might overwork themselves and their team to meet unrealistic deadlines, creating stress and reducing team morale. However, by focusing on progress rather than perfection, they could foster a healthier work environment, leading to sustainable success.

Unique Challenges

Finally, it's essential to acknowledge that each of us faces unique challenges not covered in this book. The key is to identify these hurdles and develop personalized strategies to overcome them.

For example, a tech professional might struggle with work-life balance due to the demanding nature of their job. By identifying this challenge, they can implement strategies such as setting boundaries, prioritizing tasks, and scheduling downtime to manage their workload effectively.

When you stop comparing yourself to others and focus on improving your personal average, you can accomplish great things. So, embrace your unique journey, learn from your challenges, and keep striving for growth one step at a time.

ACTIVITY: Exploring and Addressing Letting Go, Seeking Perfection, and Unique Challenges

This activity aims to help you delve into the concepts of letting go, seeking perfection, and identifying your unique challenges. By understanding these aspects, you will be better equipped to overcome hurdles and achieve your personal and professional goals.

Instructions:

- Letting Go: Reflect on what letting go means to you. Identify three or more elements in your professional and personal life you need to let go. Explain why you chose these items and how it will benefit you to release them.
- Seeking Perfection: Rate your need to be perfect and please others on a scale of 1 (low) to 10 (high). Interview two trusted individuals in your life and ask them to score your need for perfection. Request reasoning and examples. Assess their feedback and identify three new ways you commit to being kinder to yourself and increasing the focus on your needs.
- Unique Challenges: Every individual faces unique challenges. List two or more challenges unique to you that this course does not cover. Describe these challenges and explain how they affect your thinking and behavior. Identify one or more resources that can help you address these challenges. Share these details with two trusted people and gauge their understanding and support. Remember to review their feedback with your coach.
- Overcoming Challenges: Reflect on your current methods of overcoming challenges. Rate your ability to overcome challenges on a scale of 1 (hard) to 10 (easy). Note that this could vary depending on the situation. Interview two trusted individuals and ask them to rate your ability to overcome challenges. Request reasoning and examples. Identify three new ways you commit to overcoming challenges.

This activity should span over four weeks, one week per step.

Upon completion, you should clearly understand what you need to let go of, how to manage your need for perfection, be mindful of your unique challenges, and have a strategy to overcome them.

Review your answers with your coach and those who helped you in this journey.

As Saint Francis de Sales said, "Have patience with all things, but chiefly have patience with yourself." You are braver, stronger, and smarter than you think. Tough times never last, but tough people do! So, keep going and remember always to begin the task of self-improvement anew each day.

Section Summary

In this section, we dove deep into challenges, understanding their nature and how they shape our personal and professional lives. We explored various types of hurdles and identified strategies to overcome them.

We began by acknowledging the power of letting go. Letting go is not about losing control; it's about freeing ourselves from the chains of past failures or outdated beliefs that hinder our progress. By letting go, we can focus our energies on actions that truly bring us joy and lead us toward our goals.

Next, we delved into the concept of seeking perfection. The pursuit of perfection often stems from a fear of criticism or disappointment. However, we learned that this quest could be self-limiting, causing unnecessary stress and preventing us from showcasing our authentic selves. By shifting our focus from perfection to progress, we can foster growth and innovation.

Finally, we recognized that each individual faces unique challenges not covered in this book. These unique challenges require personalized strategies. By identifying these hurdles and developing tailored tactics to conquer them, we empower ourselves to navigate any obstacles that come our way.

Throughout this exploration of challenges, we engaged in reflective activities, sought feedback from trusted individuals, and developed actionable plans to overcome these hurdles. These exercises provided us with valuable insights into our behaviors and thought patterns, enabling us to cultivate resilience and agility.

In summary, challenges are an inherent part of life. They test our strength, patience, and adaptability. However, with the right mindset and strategies, we can turn these challenges into opportunities for growth and self-improvement. Remember, it's not about avoiding challenges but learning to navigate them effectively that sets us on the path to success.

RISING ACTION SUMMARY: NAVIGATING CIRCUMSTANCES, CONFLICTS, AND CHALLENGES

In the rising action of our narrative, we journeyed through the complexities of circumstances, conflicts, and challenges. This chapter aimed to equip us with the tools and insights to turn these seemingly daunting aspects into stepping stones for growth and success.

We began by exploring circumstances. We acknowledged that while we cannot control the situations we find ourselves in, we can control our reactions. We learned that we could make the most out of any situation by shifting our perspective and seeing circumstances as opportunities rather than obstacles. Whether it's a sudden change at work or an unforeseen personal event, we now have the mindset to respond effectively and proactively.

Next, we navigated the stormy seas of conflicts. Conflicts, whether interpersonal or internal, are an inevitable part of life. However, we discovered that they don't have to be destructive. We can resolve disputes to strengthen relationships and promote personal growth through open communication, empathy, and understanding. From disagreements with colleagues to inner struggles with self-doubt, we now possess the skills to handle conflicts gracefully and assertively.

Finally, we tackled challenges. These are the trials that test our strength, patience, and resilience. We examined the importance of letting go, the pitfalls of seeking perfection, and the need to identify and address unique challenges. Through reflective activities and feedback from trusted individuals, we learned to see challenges not as barriers but as catalysts for improvement and innovation.

In summary, this chapter guided us through the rising action of our narrative, arming us with the knowledge and strategies to navigate circumstances, conflicts, and challenges. With these tools, we are better prepared to face whatever comes our way and continue our journey toward personal and professional fulfillment. As we move forward, let's remember that it's not the absence of difficulties but our response to them that defines our path to success.

CHAPTER 3

Climax – The Excitement Of Building Career Development Skills

Welcome to the thrilling climax of our journey—the point where all our explorations and revelations converge, catapulting us into an exciting phase of personal and professional evolution. In this chapter, we will navigate the exhilarating world of career development skills, focusing on three pivotal areas: aiming at your North Star, skills building, and storytelling.

The first section, Aim at Your North Star, encourages you to define and uphold your values, even when confronted with challenging situations. It's about finding your own compass—your guiding light that keeps you rooted in your principles while steering you toward your goals. We'll delve into the power of a positive "no" and the art of asking questions. You'll also learn about building your tribe, drawing wisdom from some of history's most influential figures, and using their strategies to create a supportive network around you.

In the second section, Skills Building, we will equip you with the tools needed to conduct effective reconnaissance, engage your audience, and master the art of sales and persuasion. This section is akin to a toolbox, filled with practical techniques to influence key stakeholders, understand your audience, and communicate effectively. From utilizing numbers for anchoring and ranges to employing the principles of psychology to prompt others to act, you'll gain a wealth of knowledge to boost your career trajectory.

Finally, in Storytelling, we'll explore the power of narratives in shaping your professional brand. You'll learn how to project plan yourself, optimizing your current goals, and saying no to distractions. You'll craft your promotion letter with the precision of a seasoned lawyer, leveraging metrics to make a compelling case.

As we embark on this climactic chapter, remember that career development isn't just about climbing the corporate ladder; it's about personal growth and fulfillment. It's about honing your skills, building meaningful relationships, and creating a narrative that reflects your values, your journey, and your aspirations. So gear up for an exciting ride as we delve into the intricate world of career development skills.

"Success is not to be pursued;
it is to be attracted by the person you become."
—Jim Rohn

SECTION 1: AIM AT YOUR NORTH STAR

Stamp and Declare Your Values for Navigating Challenges at Work

Do you find yourself in difficult situations where you're asked to do something that doesn't align with your values?

It's essential to write down and declare your values so that when these tough decisions arise, you already have a guide to help you through them. Leverage your values as a beacon of light for your actions and decisions.

When faced with a situation that conflicts with your values, consider exercising the power of a positive no. William Ury, the author of *The Power of a Positive No*, teaches us that a positive no is an assertive way to say "no" to something you don't want to do while communicating what you can and are willing to provide or agree to. For example, if you are asked to stay late when that conflicts with your family commitments, a positive no could sound like this:

"I understand your need and appreciate the request. Unfortunately, I can't stay late today because I have other commitments. However, I'd happily work on the project first thing in the morning."

The power of the positive no is that it communicates your needs and boundaries while showing respect for the other person. It can help you stay true to yourself and your values without compromising relationships with others. So, the next time you find yourself in a tricky situation that conflicts with your values, use the power of a positive no to stay true to yourself.

Leverage this tool as a way not just to survive difficult situations but also as an opportunity to thrive. As always, trust your instincts and do whatever feels best for you! Good luck!

NAVIGATING WORKPLACE CHALLENGES: EFFECTIVE DIALOGUE STRATEGIES IN ACTION

These scripts provide a respectful and constructive way to assert boundaries while maintaining positive relationships at work. Let these examples be a springboard for you to craft your customized responses.

1. **Project Overload**
    - Boss: "Could you handle Project X as well?"
    - You: "I appreciate that Project X is important (Yes). I'm fully committed to Projects Y and Z, and I wouldn't want any quality to suffer (No). Could we perhaps delegate it to John who has shown interest in this area? Or maybe I could take it up once my current projects end (Yes)?"
2. **Unrealistic Deadlines**
    - Client: "We need this done by tomorrow."
    - You: "I understand the urgency (Yes). However, to ensure high-quality work, we need at least three days for this task (No). Could we set the deadline for the day after tomorrow instead (Yes)?"
3. **Scope Creep**
    - Client: "Could you also add feature X to the project?"
    - You: "I see how feature X can be beneficial (Yes). Our initial agreement was for the current scope, and adding tasks may affect the timeline and budget (No). If you need additional features, let's discuss a new contract for these extra tasks (Yes)."
4. **Unpaid Overtime**
    - Manager: "We'll need you to work late again tonight."
    - You: "I understand there's a lot of work to do (Yes). However, my work-life balance is suffering due to frequent late hours without compensation (No). Can we discuss a fair

overtime compensation plan or reevaluate the workload (Yes)?"

5. **Unwanted Task**

   - Colleague: "Can you handle the graphic design for this project?"
   - You: "I'm always ready to contribute to the team (Yes). However, graphic design is not my area of expertise (No). Maybe we could bring this up with our manager or ask Jane, who has a knack for design (Yes)?"

6. **Office Gossip**

   - Colleague: "Did you hear about what happened to John?"
   - You: "It sounds like you're concerned about John (Yes). I'm not comfortable gossiping about colleagues (No). Let's focus on our project instead (Yes)."

7. **Last-Minute Meetings**

   - Manager: "Can you attend a meeting in 10 minutes?"
   - You: "I understand the need for meetings (Yes). However, last-minute meetings can disrupt my work schedule (No). Could we set meetings at least a day in advance (Yes)?"

8. **Working on Weekends**

   - Boss: "We need you to work this weekend."
   - You: "I see there's a lot of work to be done (Yes). But I've committed this weekend to personal time (No). I can put in extra hours during weekdays if that helps (Yes)."

9. Taking on Someone Else's Work

- Manager: "Can you finish Tom's report? He didn't complete it."
- You: "I understand the importance of completing the report (Yes). However, it's not fair for me to take on Tom's tasks (No). Maybe we should speak to Tom or find another solution (Yes)?"

10. Unsuitable Role

- HR: "We'd like to offer you the position of Senior Analyst."
- You: "I'm truly grateful for the offer (Yes). However, the role doesn't align with my career goals (No). I'm more interested in a position related to project management (Yes)."

11. Unwanted Transfer

- HR: "We're transferring you to our branch in another city."
- You: "I appreciate the opportunity (Yes). However, I have some personal reasons for not wanting to move (No). Is there a possibility I could contribute from my current location (Yes)?"

12. Frequent Business Trips

- Boss: "You'll be traveling to client sites every week."
- You: "I understand the importance of face-to-face meetings with clients (Yes). However, frequent travel is affecting my family time (No). Could we explore alternate methods like video conferencing for some meetings (Yes)?"

13. Micromanagement

- Tech Lead: "I want to watch how you are incorporating all the changes I taught you."
- You: "I appreciate your guidance (Yes). However, constant oversight makes it difficult for me to work effectively (No). Could we try regular check-ins instead (Yes)?"

14. Excessive Work Events

- Colleague: "We're planning another after-work event this week."
- You: "It's great that we have so many team-building activities (Yes). But it's tough for me to attend frequent after-hours events (No). Perhaps we could limit them to once a month (Yes)?"

15. Salary Negotiation

- HR: "We're offering you a salary of X."
- You: "Thank you for the offer (Yes). Based on market rates and my skills, I was expecting a salary of Y (No). Can we negotiate this (Yes)?"

16. Inefficient Processes

- Tech Lead: "Stick to the process. This is how we do things here."
- You: "I see how our current process is familiar and has been working for us (Yes). However, it seems outdated and might be slowing us down (No). Maybe we could consider adopting newer, more efficient methods (Yes)?"

17. Office Politics

- Colleague: "Can you believe what management did?"
- You: "It sounds like you're upset about the situation (Yes). I prefer to stay neutral and avoid office politics (No). Let's focus on our work and how we can contribute positively (Yes)."

18. Extra Training

- Manager: "There's a training session this Saturday."
- You: "I value continuous learning (Yes). But my weekends are set aside for personal activities (No). Could we arrange for the training to be conducted during office hours (Yes)?"

19. Sharing Personal Information

- Colleague: "So, are you dating anyone?"
- You: "It's nice of you to take an interest in my life (Yes). I prefer not to share too much personal information at work (No). How about that project we're working on (Yes)?"

20. Unfair Criticism

- Tech Lead: "You always mess up the code check-in."
- You: "I appreciate your feedback (Yes). However, some of the criticism seems unjust (No). Can we schedule a meeting to discuss this constructively (Yes)?"

Show understanding and empathy, set clear boundaries, and offer a constructive alternative or solution, ensuring a win-win outcome while maintaining professional relationships.

ACTIVITY: Uncovering and Documenting Your Core Values

As a technology professional, you are likely always looking for ways to optimize efficiency and get ahead of the competition. But have you ever stopped to think about your core values? What motivates and drives you? By taking some time to uncover and document your core values, you can set yourself up for success both professionally and personally. But how do you uncover and document your values? Here's how. Grab your notebook and pen and write away!

1. Understand what values are important to you and write them down.
2. Determine how your values show up in your life and work.
3. Create a "personal value statement."
4. Live by your values every day.

By following these four steps, you can uncover and begin living by your core values, which will lead to a more fulfilling life overall. Let's dive in and take as much time as we need!

1. Identify What Matters Most

Values can deeply impact our lives. They shape our thoughts, feelings, and actions. They discover who we are as people and what's important to us while impacting how we interact with others in our communities. Values are embedded in qualities from culture, places, relationships, and personal experiences—so everyone will be unique. If thoughtfully documented, your values become like a compass for your life, helping guide decisions, goals, and aspirations. A value system evolves over time as you discover yourself and what matters most to you: it's about learning more about who you are at your core. Understanding your values is key to self-knowledge and can help you understand yourself more deeply.

The first step in discovering and documenting your core values is identifying what matters most. Think about the people, places, or ideas most important to you throughout your life and any goals or aspirations you have for yourself. These concepts can provide critical insights into what truly matters to you at a deeper level.

2. Ask Yourself Questions

Taking the time to ask yourself questions and understand your values is an essential but often overlooked step in personal growth. If you're unsure of your values or how they fit into your life goals, ask yourself more specific questions on topics such as relationships and careers. Consider what you want from each relationship and ask forward-looking questions, such as "How can I measure success?" or "What will my job provide for me?". Reflecting on these deeper concerns can help establish your decisions moving forward. Ultimately, it's essential to be aware of your values and remain conscious of how they shape your daily activities—it's a key factor in leading a meaningful and fulfilling life!

Once you identified some things that matter most, it's time to dig deeper by asking yourself why they are essential. Why did these specific people, places, or ideas make such a huge impact on your life? What qualities about them made them so meaningful? Answering these questions can help clarify the values that truly matter to you and guide how you want to live your life from now on.

3. Brainstorm Ideas and Refine Your List of Values

Core values are essential to understanding who you are and what matters most to you. To uncover these core values, start by reflecting on your beliefs and reasons for why they are important. Making a core personal value statement can help bring clarity and focus, allowing you to make informed decisions aligning with those values.

Once you have identified the core values, develop a slogan that easily explains them.

Now it's time for the fun part—coming up with ideas! Brainstorm as many words or phrases as possible related to the things that matter most in your life (i.e., loyalty, respect, community). When complete, review this list of words/phrases and refine it into a more concise set of core values (around seven to ten) that are meaningful and actionable for both yourself personally and professionally.

4. Write It Down and Share It With Others

Sweeping change often starts with small steps, and documenting your findings is the first one. The knowledge that comes from documenting your discoveries can create a definitive list of things you should do differently. Once you identify and refine your list of values, write them down in an easy-to-reference document or place where your trusted circle can access them as well. Then try experimenting with different strategies. This can create great insight into which ones have the most impact. After doing this, create a review of what worked, what didn't, and what needs to be updated for the next time. Finally, share this information with those who need it so everyone can make real progress. Taking these simple steps to document your findings and put them into action will create clear paths toward lasting success in any organization or endeavor.

Uncovering and documenting our core values gives us a greater sense of purpose, allowing us to make more informed decisions based on what we believe in at a deeper level—both personally and professionally. By taking the time to identify our true "North Star," we can ensure that our actions align with our beliefs moving forward—no matter what comes our way.

If you can identify your values and live by them each day, you will create a foundation for professional and personal success. You can make better decisions, attract the right people, fulfill your potential, and find true happiness. So, where do you begin? Start today by brainstorming what values are important to you. Once you understand that, start living by those values every day in small ways. Be consistent with it; soon enough, it will become second nature. So, take some time today, think deeply, uncover, document, share, and start living aligned!

ACTIVITY: Develop a Not-To-Do List

If you're a tech professional, chances are you already have your to-do list nailed down. But to be as productive and successful as possible, it's time to start creating a Not-To-Do list. A Not-To-Do list is just as important—if not more so—than the typical "what needs done" checklist. It helps prioritize the projects that matter and reduces wasted time on tasks that don't contribute meaningfully. Without one, it can quickly become too easy for even well-meaning tech professionals like us to become sidetracked by less urgent demands or inefficient behaviors. Apply the following steps to establish your Not-To-Do list and win everywhere. At this point, get your notebook and pen.

1. Start

Start by reviewing your current daily and weekly tasks and activities. Take time to consider what you are currently doing, why you are doing it, and how much time it takes away from the things that truly matter to you. If a task or activity does not contribute to your goals or personal satisfaction in any meaningful way, it may be a good idea to cross it off your list.

2. Identify

Identify the boundaries that you want to set for yourself. Ask yourself questions, such as "What am I willing or not willing to sacrifice to make time for my priorities?" and "Is this activity worth my precious time?" Pinpoint where your values lie, and determine if an activity fits within those boundaries.

3. Create

Once you identified the items on your list that don't fit your values or goals, create a "Not-To-Do " list. This list should include everything that is not essential for you to do on a daily/weekly basis for you to achieve your goals and live according to your values. It can also include activities or tasks that many people might see as important but are not essential for you to do right now (e.g., cleaning out the garage).

4. Prioritize

Prioritize the items on your "Not-To-Do" list so that they are in order of importance and relevance for achieving your main objectives and living according to your values. This will help ensure that these items are removed from your weekly schedule, rather than being pushed aside without being actively addressed.

5. Follow

Stick with the plan! Make sure not only to cross off these tasks from your weekly schedule but also avoid them like the plague—actively choosing not to do them even when presented with the opportunity (which can be difficult if family members, friends, etc. are involved). This will help ensure that you free up more time for yourself while helping solidify better habits over time—ones that place importance on living according to one's core values rather than constantly succumbing to peer pressure (no matter who it is coming from!)

Learn from Carlos how he said no to succeed at work and not compromise his friendship simultaneously:

It was a busy Monday morning, and I had an important presentation at work. My promotion depended on it, so I knew preparing for the presentation was my top priority. As if that wasn't enough stress, my friend called me and asked for help moving over the weekend.

I felt torn between maintaining our friendship and investing in my career growth. If I said yes to helping him move, then there would be no time for me to focus on getting ready for the big meeting at work. On the other hand, saying no would mean letting down someone who has always been there when I needed them most.

After some thought, I decided to talk to my friend about the situation honestly. After all, he deserved an explanation for why I couldn't help him move this weekend, not just a cold "no."

I told him that while his request meant a lot to me and that nothing could take away from our strong bond of friendship—even something like this—it was imperative that I prioritize completing this project before anything else right now or else risk losing out on a great opportunity professionally speaking.

He understood where I was coming from, and we agreed that instead of helping with the move itself—which clearly wouldn't be possible due to timing constraints—I'd come by later in the week after his furniture arrived and lend a hand arranging everything inside his new apartment properly! This way, we can both get what we need: he'll get help with moving into his place, and I'll have plenty of time left over during the day/evening hours leading up until Saturday night (when my presentation is due) to prepare properly without having any distractions or extra commitments weighing me down!

Steps Carlos took:

Yes – to prioritize his career.

No – to help his friend move.

Yes – to help his friend arrange his furniture instead.

Why does this method work so well for building better habits?

Habits are hard to change. That's why so many of us end up repeating old, unhealthy behaviors. This method for building better habits is so effective; it works by focusing on the small and achievable steps to reach our goals. This method helps lay the groundwork for sustainable, positive change. It allows us to track progress, adjust as needed, and be held accountable for each step. Increased motivation and confidence in our ability to change make us more likely to achieve lasting success.

ACTIVITY: Learn to Say No and Create Win-Win Situations

1. Read the book *Power of a Positive No* by William Ury. When time is limited, you can access the book summary online.
2. Write five or more (real or hypothetical) examples of your Yes-No-Yes script. This can be at work, home, friends, train station, coffee shop, school, etc.
3. With the help of someone you trust or in front of a mirror, practice delivering your Yes-No-Yes scripts.

To say no is to protect your time and mental space for things that matter. It's an act of self-compassion, allowing you to focus on your goals and priorities. Learning how to say no politely but firmly is a skill worth developing. The benefits are many—better work/life balance, improved relationships, less stress, and more opportunities to do things you love. If you're ready to start saying no more often, here's a workflow example you can use or customize to your benefit today. The key is to be clear about what you're saying no to, get comfortable with saying no out loud, and practice regularly until it feels natural.

The Art of Asking Questions

Asking questions can be a powerful tool for achieving your goals. Whether you are a design engineer or a leader, effective questioning can help you gain information, investigate, provide safety and security, trigger thinking, invoke philosophy, guide, calm, and develop understanding. Most often, we think of asking questions to obtain answers. But what if you asked the right questions for the best possible outcome?

Harnessing the Power of Questions

Asking questions can also be used to agree with someone else's point of view. By asking the right questions, you can acquire closure or peace about an issue that may have been previously unresolved. For example, "What do you think would be the best approach for this project?" Asking these types of questions shows that you are actively listening and are interested in their opinion.

Questions can also be used to prepare yourself for an upcoming activity or event. For example, "What do I need to consider before I present my idea?" Thinking ahead through questioning is a great way to ensure you are as prepared as possible before acting on any task.

Questions can also be used to defend yourself against objections or criticism. For example, "How did our team handle that situation differently than other teams?" Asking this type of question allows you to showcase your team's effectiveness and successes without putting yourself in a defensive position.

Questions can also influence another person's opinion on a subject. For example, "Why do we need more resources when we already have enough staff?" This type of question forces them to think critically about why they need additional resources and provides an opportunity for discussion.

Lastly, asking questions effectively convinces someone of something that may not have been previously considered. For example, "What could we do differently if we wanted our project to have greater results?" Presenting this scenario encourages people to look at things from different perspectives, which may lead them to find solutions that were not previously considered.

Questioning can be a powerful tool for establishing a safe and secure environment. For instance, consider the question, "How do you feel about this decision?" This simple inquiry acknowledges the other person's emotions, making them feel seen and heard. It creates a safe space for them to share their thoughts and feelings without fear of judgment or criticism.

Questions can shield you when you're feeling vulnerable or under pressure. For example, asking yourself, "What steps do I need to take to ensure my well-being in this situation?" encourages self-reflection and helps you devise a plan to protect your mental and emotional health.

Questions can be your defense mechanism in the face of criticism or objection. Instead of responding defensively, ask, "Can you help me understand why you think our approach was less effective?" This defuses tension and opens up a dialogue for constructive feedback and learning.

My colleague and friend Andy Betts, founder of ICONDA Solutions, conducted a remarkable study and captured the different types of questions and their valuable use cases. Let's review his teachings next.[20]

Understanding the Types of Questions

As an artist uses different brushes to create a masterpiece, we use different types of questions to paint a clearer picture of our understanding. Each question type serves a unique purpose and can lead us to the desired outcome when used effectively. Here are seven types of questions that you can leverage in your professional journey:[21]

- Umbrella Questions: These broad, overarching questions open up a conversation. They allow for various responses, setting the stage for deeper exploration. Think of them as opening the doors to a vast library of knowledge. For example, "What are the key objectives of this project?" or "Can you give me an overview of our team's responsibilities?"
- Clarification Questions: These questions help you delve deeper into specific details. They remove ambiguity and ensure a clear understanding of the subject matter. Like using a magnifying glass, they bring clarity to the most minor details. For example, "Could you explain the function of this code?" or "What does this metric represent in our report?"
- Opinion Questions: These are designed to elicit personal viewpoints or suggestions. They foster engagement and show respect for the other person's thoughts. Use them to build relationships and gain diverse perspectives. For example, "What is your opinion on this design?" or "How would you approach this problem?"
- Open Questions: Open questions invite expansive answers. They stimulate thought and encourage a full, meaningful answer using the subject's own knowledge and/or feelings. In essence, they're tickets to an open-ended conversation. For ex-

ample, "How has this software changed the dynamics of our work?" or "What are the potential impacts of this decision?"

- Closed Questions: Closed questions are your fact-checkers. They usually require a simple yes-or-no answer or factual information. They're the nuts and bolts holding your understanding together. For example, "Is the client meeting scheduled for Monday?" or "Did the update fix the bug?"
- Curiosity Questions: These are driven by genuine interest and a desire to learn more. They foster innovation and creativity. Like a curious child exploring the world, they lead you to new insights and discoveries. For example, "What inspired the creation of this feature?" or "What's the story behind this company tradition?"
- Important Questions: These questions address critical issues or concerns. They help prioritize tasks and make informed decisions. They're like the lighthouse guiding your ship amidst a storm. For example, "What are the potential risks of this approach?" or "How does this align with our company's mission?"

Each of these questions has its unique value and intention. Understanding when and how to use them can drive effective communication, encourage learning, and foster collaboration.

Remember, the art of asking questions isn't just about getting answers. It's about engaging in meaningful conversations, building relationships, and fostering an environment of continuous learning. So, go ahead, ask away, and watch as the world opens up to you in ways you never imagined.

How to Ask Questions that Lead to Impactful Outcomes

The first step in asking questions that lead to impactful outcomes is defining the outcome you hope to achieve. What do you want to accomplish? What change do you want to see? Once you have a clear idea of the outcome you hope to achieve, you can formulate your questions.

The next step is to identify the stakeholders in the situation. Who will be affected by the outcome of the situation? Who has a vested interest in the success or failure of the situation? Once you identify the stakeholders, you can formulate questions to help you understand their needs and concerns.

The third step is to gather information about the situation. What are the facts of the case? What is the history of the situation? What are the current conditions? The more information you have about a situation, the better equipped you are to ask questions that lead to impactful outcomes.

Once you gather information about the situation, you can formulate questions to help you achieve your desired outcome. It is important to avoid yes-or-no questions when asking questions, as they will not provide you with the information you need to make an impactful decision. Instead, focus on open-ended questions, allowing stakeholders to share their thoughts and feelings about the situation.

It is essential to listen carefully when stakeholders are answering your questions. Pay attention to their body language and tone of voice as these can provide clues about their thoughts and feelings. In addition, take time to process what they say before responding. This will show that you are truly interested in their input and value their opinion.

If there are parts of a stakeholder's answer you do not understand, or if you want more information about a particular issue, ask follow-up questions. This shows you are engaged in the conversation and committed to finding a resolution meeting everyone's needs.

Taking notes during a conversation can also help to prevent question stacking. This is because it will allow you to keep track of what has been said and ensure that your questions are relevant to the conversation. Additionally, taking notes can help you remember important details necessary for follow-up questions later.

After speaking with all of the stakeholders, it is essential to summarize what you heard. This will help ensure that everyone is on the same page and that there is a shared understanding of the situation. It will also help prevent misunderstandings or miscommunication down the road.

Consider this handy checklist when asking questions:

- Do your research ahead of time: Before you ask a question, it's imperative to do your research and ensure you clearly understand the topic. This will help you to ask more informed and specific questions. Additionally, it's advantageous to be familiar with the company or organization to which you're asking the question. This will help to ensure that your question is relevant and on-topic.
- Make sure your question is clear: When you ask a question, make sure it is clear and concise. Avoid asking questions that are too broad or general, as this can be frustrating for the person trying to answer them. Additionally, try to avoid asking multiple questions at once. This can be overwhelming and make it difficult for the person to provide a thorough answer.
- Make sure your questions are relevant to the discussion: It's critical to make sure that your questions are relevant to the discussion at hand. If they're not, you risk derailing the conversation and wasting everyone's time.
- Ask clarifying questions if you're not sure you understand something: If you're unsure you understand something, asking a clarifying question is perfectly acceptable. This will help ensure everyone is on the same page and that no one is left in the dark.

- Be respectful: When you ask a question, always respect the person you are asking. This includes refraining from interrupting them while they are speaking and listening carefully to their answer. Additionally, avoid asking personal questions or making assumptions about someone's qualifications or experience.
- Be aware of your body language: Your body language can impact how others perceive your question. It's essential to maintain eye contact when you're asking a question and to avoid crossing your arms or legs, as this can make you appear closed off or uninterested in the answer. Additionally, try to speak in a clear and confident voice so that you come across as sincere and interested in what the other person says.

Conversely, here are elements to consider as part of a To Don't list:

- Avoid question stacking: Asking multiple questions at once can often lead to confusion and cause the person you are speaking with to lose track of what was said. Therefore, it is important to ask one question at a time and wait for a response before asking another question. This will help to keep the conversation flowing smoothly and prevent any misunderstandings from occurring.
- Avoid yes/no questions: Yes/no questions can often be perceived as confrontational or challenging. Instead of asking a yes/no question, try phrasing your question in a way that allows for a more detailed answer. For example, instead of asking, "Do you have experience with X?" you could ask, "What experience do you have with X?"
- Avoid asking loaded questions: Loaded questions can't be answered with a simple yes or no. They're often used to trap people into admitting something that they may not want to

admit. Loaded questions should be avoided as they can often lead to arguments and hurt feelings.

- Avoid asking personal questions: Personal questions are those that are too intimate or personal for the situation. They can make people uncomfortable and often lead to awkwardness and tension. If you're not sure whether a question is too personal, err on the side of caution and avoid asking it altogether.

ACTIVITY: The Art of Questioning in a Business Context

This activity is designed to help you observe, analyze, and improve the dynamics of business meetings through the power of questioning. You'll be studying people's behaviors, interactions, and communication styles and harnessing your observations to develop strategies for more productive and efficient meetings.

Materials Required:

- A notebook that you've been using throughout this book
- Pen or pencil
- A digital recording device (optional)

Instructions:

- Identify the Observation Context: Choose a business meeting you'll observe. Note down details like its type, participants, duration, etc.
- Establish Specific Goals: Define what behaviors you aim to observe and analyze. Are there any particular aspects or control factors you need to consider?
- Gather the Necessary Tools: Decide on the tools you'll use for recording or note-taking during the meeting.

- Develop a Timeline: Determine the time you'll spend observing the meeting and analyzing your findings.
- Set up an Experiment Plan: Strategize how you'll observe the meeting and record your findings. List any specific questions you might need to ask or answer before starting.
- Compose Questions: Draft 15-20 questions you'll ask yourself during or after the experiment to observe behavior within the meeting. These questions should aim to develop your skills in asking different types of questions, enhance your observation skills, and encourage listening before speaking.
- Record the Results: During the meeting, meticulously record your observations and findings on a digital device or in your notebook.
- Analyze the Results: After the meeting, take some time to dissect your findings. Look for patterns, trends, or areas that need further exploration.
- Publish the Findings: Once you analyze the data, write a summary of your experiment in your notebook. Conclude with recommendations for improving behavior in business meetings based on your findings.

These questions aim to help you analyze the meeting's communication dynamics, leadership styles, and decision-making processes:

1. What conversations are taking place?
2. How do people respond to each other during the meeting?
3. Are there any patterns in how people interact with one another?
4. Is there a clear leader of the meeting, or is it more collaborative?
5. How do individuals take turns speaking?
6. Are any conversations dominated by one person or a small group?

7. What body language is being used to communicate?
8. Is there an appropriate amount of eye contact between participants?
9. How are different types of communication (verbal, nonverbal, etc.) used during the meeting?
10. How are conflicts handled within the group?
11. Are there any areas that need improvement regarding communication among participants?
12. Is there a general feeling of consensus or agreement among attendees?
13. Does anyone seem reluctant to participate in discussions?
14. Are decisions made quickly, or is there a lot of deliberation?
15. How is the meeting managed overall?
16. What strategies can be implemented to increase engagement and participation during the meeting?
17. Are there any topics that could have been discussed more thoroughly?
18. Could conversations have been framed differently to achieve a better outcome?
19. Are any activities or tasks overmanaged?
20. What changes can be made to ensure that meetings are productive and efficient?

This self-paced activity depends on the duration of the meeting you observe. However, try to set aside some time immediately after the meeting to record and analyze your observations while they're still fresh.

By completing this activity, you'll gain insights into the behaviors, interactions, and communication styles prevalent in business meetings. You'll also develop questioning and observational skills, enabling you to contribute more effectively to future meetings. The ultimate goal is to use these insights to devise strategies to enhance engagement, collaboration, and productivity in business meetings.

Asking questions is an essential skill for success in both professional and personal life. Whether it's to better understand a situation or pursue desired results, asking the right questions can help you achieve your goals. This chapter explored how asking questions can help you achieve your goals, the various intentions, and types of questions to consider expanding your versatility of asking pragmatic questions, and how getting others to say "no" is an effective way. You'll be able to gain valuable insight into any situation and increase your chances of achieving your desired outcomes faster than ever before.

Building Your Tribe

It is said that it takes a village to raise a child. Similarly, it takes a tribe to help each other professionally grow. As we know, the world is constantly changing, and technology is ever-evolving. To stay ahead of the curve, it is essential to have a supportive network of people who can help us learn and grow. A tribe provides its members with a sense of belonging and connection and opportunities for personal and professional development.

Belonging to a tribe is an important concept that brings us together in more ways than one. It provides us with belonging and support. It can help us stand up together, united under one cause when times are tough. It gives us a sense of belonging and comes with strength in numbers. We can use this collective strength to help bridge gaps and learn from each other's differences, seeing beyond our blind spots. Through belonging to a tribe, it becomes easier for us to offer and be offered help when we need it most. As Nelson Mandela said, "A nation should not be judged by how it treats its highest citizens, but its lowest ones."[22] Belonging to a tribe reinforces this notion—lending our support as one in times of need.

Every search for a tribe should begin with a search for values that resonate with you. Ones that provide the moral spine supporting your search quest. Once you've found these, you can then search out and follow those who may resemble and lead the tribe in which you hope to be part. Remember, within a larger community of like-minded individuals, true strength lies as it compounds the power of each leader and professional. However, it all hinges on finding a group whose principles make your heart soar.

To be part of a tribe is to be a part of something bigger and benefit from moral and ethical values that expand one's boundaries. In the diversity of opinion, we must learn to adapt and grow our understanding to enable solutions for more tremendous success than could ever be imagined alone. Every member is a fallback by having others around to support with different perspectives as bridges for new ideas and outcomes. Herein lies the sacred Zen between traditional customs and modernized ideals. Together, we remember what it means to walk hand in hand toward a future filled with promise and hope.

Joining a tribe offers immense opportunities to multiply, invite, grow, and learn. We can expand, improve our technical prowess, share our professional expertise, and gain exposure to new perspectives as a group. When participating as an individual within the tribe, one can take on the roles of manager and mentor and benefit from being managed and mentored. These unique interactions among members are a powerful tool that opens us up to different possibilities and takes us beyond what any single person could ever do alone.

ACTIVITY: Building Your Tribe at Work

This activity is designed to guide you through joining or forming a tribe at work. A tribe is a group of like-minded individuals sharing common interests and values who offer mutual support to achieve professional goals. Whether you're looking to connect with colleagues in similar roles or seeking diverse perspectives outside your immediate team, building your tribe can be a gamechanger for your career. Get your notebook and pen or pencil.

Instructions:

STEP 1: Define your purpose and values

- What is your purpose for creating a tribe at work? Start with outcomes, then methods, and, lastly, timelines.
- What shared values do you want your tribe to embody? Go deep, 20 to 30 layers deep. This is intended to squeeze quality values. Trim the final list to the top seven to nine values.
- Write a brief mission statement that summarizes your purpose and values.

STEP 2: Identify potential tribe members

- Who are the colleagues or peers who share your interests and values? Document a list.
- What skills and experiences do they bring to the table? Identify categories and utilize this to add variety to the tribe.
- Reach out to these individuals and invite them to join your tribe. Draft a welcome note and be ready to share your mission, purpose, and values.

STEP 3: Establish guidelines and structure

- What are the roles and responsibilities of each member?
- How often will you meet and communicate?
- What communication channels will you use?
- Define your group norms and expectations for behavior.

STEP 4: Foster trust and communication

- Establish trust by documenting and then sharing your own experiences and perspectives.
- Encourage open communication and active listening. Share and agree on how this is achieved.
- Create opportunities for collaboration and teamwork. Write these down and revise them with the tribe.

STEP 5: Set goals and success criteria

- What are the specific goals you want to achieve as a tribe?
- How will you measure success?
- Define clear metrics and timelines for achieving your goals.
- Step 6: Document your plan and progress
- Create a document outlining your purpose, values, guidelines, and goals.
- Record minutes from your meetings and decisions made.
- Track progress toward your goals and adjust your plan accordingly.

STEP 7: Evaluate the outcomes

- Did you achieve your goals? Why or why not?
- How did being part of a tribe affect your personal and professional growth?
- Use feedback from members to improve the group's structure and processes.

As the tribe leader, balancing formal, detail-oriented discussions and softer, more casual conversations is crucial. Here are some guidelines on how to switch between these two types of communication:

- Understand Your Tribe Members: Knowing your tribe members' personalities, communication styles, and preferences will help you determine the right approach. Some people may prefer more formal, structured conversations, while others might thrive in a relaxed, casual atmosphere.
- Set the Tone: The type of conversation often depends on the situation or context. For formal discussions, such as goal setting, project planning, or performance reviews, it's imperative to be clear, concise, and professional. On the other hand, casual conversations are great for brainstorming sessions, team bonding activities, or just catching up on a personal level.
- Be Adaptable: Learn to read the room and adjust your communication style accordingly. Lightening the mood with a joke or personal anecdote might be helpful if a meeting becomes too tense or rigid. Conversely, if a casual conversation veers off topic, gently steer it back with a focused question or statement.
- Encourage Open Dialogue: Create an environment where all communication styles are valued and respected. Encourage tribe members to express their ideas and opinions in a formal presentation or a casual chat.
- Lead by Example: Show your tribe members that switching between different communication styles is okay. Be transparent about your expectations and demonstrate both formal and informal communication in your interactions.

The goal is to create a comfortable space where everyone feels heard and valued. Balancing formal and informal conversations can help ensure that all tribe members are on the same page, fostering camaraderie and mutual respect.

Counter Your Echo Chamber: Avoid Pitfalls When Building Your Tribe

Creating a tribe at work can be a powerful tool for collaboration, innovation, and support. However, it's crucial to avoid forming an echo chamber, where only similar ideas and perspectives are shared, leading to groupthink and discouraging diversity of thought.

Dealing with difficult stakeholders can be challenging, but it's essential to understand that effective communication and active listening can make all the difference. One of the most valuable tips when approaching these situations is to practice beforehand. This can help you build confidence and develop a strategy that works best for you. Listen carefully to what the stakeholder says when it's time to communicate. Try understanding their concerns and needs and showing that you're genuinely interested in finding win-win outcomes. Acting upon their feedback and ideas can help establish a positive rapport. Patience is vital when dealing with difficult stakeholders. By taking the time to communicate and actively listen, you can foster a constructive relationship and achieve successful outcomes.

When two parties conflict, it can be easy to become entrenched in one's perspective and forget about the other's point of view. However, empathy can be a powerful tool in bridging this gap. Creating a safe environment where both parties feel heard and understood is essential to use empathy effectively. This can involve active listening and asking questions to understand the other's perspective better. It's also important to be firm where necessary, setting boundaries and expectations for constructive communication. By employing empathy in this way, you can create a space where both parties feel empowered and seen. In the end, you may win them over to your side.

In the fast-paced world of work, it's easy to jump right in and get things done quickly. However, being patient and getting to know the people you're working with can make all the difference in the long run. Asking questions and seeking to understand their perspectives can lead to more effective collaboration and better outcomes for everyone involved. It's also important to help others reveal their fallacies safely rather than immediately pointing out their mistakes. Building strong relationships and creating a safe feedback space can cultivate a more positive and productive work environment.

Practicing healthy debate, negotiation, and compromise is essential when reaching a consensus. Healthy debate allows everyone to voice their ideas and concerns, ensuring that all viewpoints are heard. Once everyone has spoken, it's crucial to differentiate between nonnegotiables and soft concerns. Nonnegotiables are the key issues that must remain unchanged for the group to move forward, while soft concerns are where the group can compromise. It's important to endorse unity throughout the process, knowing that while everyone may not get everything they wanted, the team will still come away with a satisfactory outcome. Ultimately, using these techniques, a group can reach a consensus that works for everyone.

In today's increasingly divided world, finding common ground is more important than ever. By focusing on shared values and interests, we can double down and amplify our common goals, creating a win-win situation where no one needs to feel defeated. It's all about recognizing that we all have similar desires and needs, regardless of our backgrounds or beliefs. Whether working toward a common environmental goal or building stronger communities, finding common ground is the key to success. We can bridge the gaps that separate us through dialogue and mutual understanding and create a bright future for all.

Taking risks and trying new strategies can be intimidating in business, but it's important not to let fear hold you back. Instead, focus on taking calculated risks that have the potential to yield great rewards. One way to do this is by experimenting with new approaches and techniques. By innovating and embracing new ideas, you can activate your business and propel it forward. Of course, not every strategy will succeed, but repeating the process and working toward your business goals are essential. Don't be afraid to step outside your comfort zone and take a chance. Who knows what opportunities may arise from trying something new?

By applying diplomacy lessons to building your tribe, you can create a supportive and collaborative network that helps you achieve your professional goals and fosters a win-win situation for everyone involved. Diplomacy is a valuable skill for you as it allows you to navigate complex situations and build strong relationships with your colleagues and stakeholders.

Section Summary

In the journey of career development, remember these key takeaways:

1. Stand firm on your values: Life will throw you curveballs, but it's not the situations we find ourselves in but how we react to them that defines us. If a situation continually conflicts with your values, it's time to act. Either change it or step away safely. A positive no is a powerful tool—say yes to your values, no to the conflict, and yes again to a solution that aligns with your principles.
2. Master the skill of asking questions: The power of a question lies not just in its answer but in its intent and the dialogue it sparks. Different types of questions serve different purposes, and understanding this can be a leadership game changer. Remember, the quality of your questions determines the quality of your life.
3. Build your tribe and beware of echo chambers: Creating a supportive work environment is like building a tribe. It fosters collaboration and drives productivity. But don't let your tribe become an echo chamber where only similar voices are

heard and diversity of thought is lost. Promote diverse perspectives, encourage open dialogue, and practice diplomacy. Always keep a finger on your tribe's pulse, adjusting as necessary to ensure a healthy dynamic.

Hold on to these nuggets of wisdom, share them, and see how they can transform your career development journey.

SECTION 2: SKILLS BUILDING

As we journey further into the exciting realm of career development skills, we're about to delve into the nitty-gritty—learning to conduct reconnaissance, engage effectively, and master the art of sales and persuasion. Remember, every step you take is a step toward becoming a better version of yourself.

In the upcoming pages, you'll uncover the power of MAP, PAGE, and Discover-Y from ICONDA Solutions[20]. These tools will help you map discussions, research your audience from their perspective, and dive deep beneath the surface of problems to reveal the needs that need to be facilitated.

You'll also learn to craft compelling two-minute messages (TMMs)[20] and two-second messages (TSMs)[20], using the power of numbers, psychology, and strategic communication to influence others and accomplish your objectives. Whether building persuasive arguments or composing concise emails, these techniques will be your guiding stars.

Finally, we'll explore the intriguing world of sales and persuasion, where everyone sells, even if they don't realize it. From understanding loss aversion and opportunity cost to leveraging the potential of novel experiences and the paradox of choice, you'll acquire the tools to navigate this landscape with confidence and finesse.

So, buckle up and get ready for an enlightening ride. This isn't just about learning new skills—it's about transforming the way you approach your career and life. Let's dive in!

Conduct Reconnaissance

The Power of MAP[20]

In engineering and technology, problem-solving is often hailed as a badge of honor. However, this strength can also become an Achilles' heel if not wielded with care. While the hammer of problem-solving is an essential tool, there are times when the delicate strokes of a paintbrush might serve better.

This section introduces you to a powerful concept developed by Andy Betts of ICONDA Solutions: MAP—My objectives, Audience context, and Plan. Rather than reacting immediately to situations, we encourage professionals to pause, step back, and create a MAP to guide their actions.

The first step involves defining your objectives. Clearly understanding what you aim to achieve is crucial, whether it's an engagement, a project, or a sales call. As the saying goes, "a problem well-defined is a problem half-solved."

Next, delve into understanding your audience. Go seven layers deep into their psyche, beyond just the surface-level needs. For instance, a product marketing engineer pushing for a new feature might be doing so to improve the product or increase sales and as a stepping stone toward a promotion. Recognizing these hidden interests can pave the way for more meaningful and productive interactions.

But how do you balance these intricacies without risking product development timelines, workplace dynamics, and other factors? This is where the final element of MAP comes into play: the Plan. With a well-defined objective and a deep understanding of your audience, you can devise a plan that accommodates these complexities while still driving toward your goals.

Every situation demands a different tool. By adopting the MAP approach, you can ensure you're equipped with the right one at the right time, transforming challenges into opportunities for growth and success.

PAGE[20]

In the modern workplace, efficiency and clarity are key. One tool that has proven effective in fostering these is PAGE—Purpose, Agenda, Goal, and Endorsement. This concept, an integral part of ICONDA Solutions' suite, is designed to streamline communication and ensure meetings are productive and purposeful.

The first step in the PAGE process is defining the Purpose. This involves clearly outlining why a meeting or initiative is necessary. It provides direction and serves as a beacon guiding all subsequent actions.

Next comes the Agenda. This is the roadmap of the meeting, outlining what will be addressed and when. It keeps discussions on track and ensures every important point is covered.

The third component is the Goal. This differs from the Purpose in that it's more specific, detailing the tangible outcomes you aim to achieve from the meeting or initiative.

Finally, we have Endorsement. This involves gaining the support of all involved parties for the Purpose, Agenda, and Goal. Endorsement ensures everyone is on the same page and fosters a sense of collective responsibility toward achieving the outlined objectives.

Now, let's bring this concept to life with a couple of relatable anecdotes from the tech world.

Consider a software development team planning a sprint. The Purpose could be to enhance the user experience of their app. The Agenda might include reviewing user feedback, brainstorming improvements, and assigning tasks. The Goal could be to reduce app loading time by 20 percent and increase user session duration by 10 percent. Before proceeding, the team lead seeks Endorsement from all team members, ensuring everyone understands and supports the plan.

Alternatively, imagine a marketing team at a tech firm planning a product launch. The Purpose is to introduce the new product to the market successfully. The Agenda might involve finalizing the press release, planning social media campaigns, and coordinating with influencers. The Goal could be to achieve 10,000 product sales in the first month. The marketing manager secures Endorsement from the team, ensuring everyone is aligned and committed to the plan.

In both cases, PAGE ensures clarity of purpose, efficiency in execution, and alignment among team members. It's a powerful tool that can transform meetings from time-consuming necessities into engines of productivity and innovation.

Often, the phrase "we had a good meeting" holds little weight. It's akin to a cloud in the sky—present but nebulous, offering no clear direction or actionable steps. However, when we employ tools like PAGE, we transform this cloud into a roadmap. We illuminate the path with Purpose, Agenda, Goal, and Endorsement. These are not just words but pillars that hold up the tower of constructive dialogue and progress. They provide clarity amidst the vagueness, ensuring that even in the absence of agreement, there is a tangible framework to guide actions and foster growth.

Discover-Y: A Deep Dive into Effective Collaboration[20]

In our professional journey, we often navigate an intricate web of interactions, expectations, and deliverables. These engagements, whether with our managers or cross-organizational stakeholders, are frequently coated with data, polished presentations, and the unavoidable politics of the workplace. Amidst this complexity, we sometimes lose sight of the true essence of human and professional interaction.

Enter Discover-Y, a tool designed to bring a multidimensional approach to accomplishing tasks at work. Picture the letter Y. At its base is you. Where the three lines intersect, that's your audience—the individual you aim to understand better to achieve specific outcomes. The two prongs extending from the intersection point toward two categories: living elements (peers, managers, direct reports, stakeholders, etc.) and non-living elements (budgets, plans, projects, commitments, politics, etc.) associated with your audience.

As you sketch out this mental map, you start creating categories of questions. One set pertains to activities, interactions, and commitments between you and your audience. Another set probes the relationships between your audience and the living elements associated with them. A third category delves into the interactions between your audience and the non-living elements.

From here, you begin to assign legends to the living entities with which your audience interacts. Whom does your audience trust? Whom do you trust? Who trusts your audience? Who trusts you? Suddenly, trust becomes a multidirectional vector, providing new layers of insight. You can introduce other vectors too, such as competence, influence, and relationship strength. This process can also be repeated for the non-living elements, adding even more depth to your understanding.

This exercise may initially seem overwhelming, but even the most complex tapestry begins with a single thread. With practice, it will become second nature. By embracing the Discover-Y approach, you become akin to a master chess player, seeing moves ahead and understanding the board. You'll be able to navigate workplace interactions with newfound clarity and strategy, devoid of manipulation or politics.

Let's illustrate this with an example. Suppose you're a project manager preparing for a critical meeting with a senior executive. At the base of the Y is you. The intersection point is the senior executive, your audience. The living elements associated with them could include their team, their peers, their manager, and perhaps crucial customers. The non-living elements might encompass their budget, ongoing projects, performance targets, and office politics.

As you start asking questions and assigning legends, you'll begin to see patterns and connections that weren't apparent before. You might discover that the senior executive trusts their team implicitly but has a strained relationship with a crucial customer. This information can help you tailor your approach, focusing on team strengths while delicately addressing customer concerns.

Discover-Y is not just a tool; it's a compass guiding you through the often-turbulent seas of professional interaction toward the shores of effective collaboration and mutual success.

ACTIVITY: Integrating MAP, PAGE, and Discover-Y

This activity is designed to help you apply the concepts of MAP, PAGE, and Discover-Y in real workplace situations. By actively doing this exercise, you'll better understand these tools and how they can enhance your professional interactions.

Materials Required:

- The notebook you've been using for this book
- A pen or pencil
- A quiet space for reflection

Instructions:

First, choose one of the five example workplace encounters provided or customize an encounter that suits your situation.

- Software Development Project: You're a project manager overseeing a new software development project. The primary intercept point on your Discover-Y mind map is the lead developer, whose expertise and commitment will be crucial to the project's success. Additional stakeholders include team members, the client for whom the software is being developed, and senior management, who expects timely and successful completion of the project.
- Product Launch: You're the product manager about to launch a new technology product. The primary intercept point on your Discover-Y mind map is the marketing manager, who will communicate the product's value proposition to customers. Additional stakeholders include the sales team, customer service representatives, and the executive leadership team.
- IT Infrastructure Upgrade: You're an IT director planning a major upgrade to your company's IT infrastructure. The primary intercept point on your Discover-Y mind map is the Chief Information Officer (CIO), whose support and approval are necessary for the upgrade to proceed. Additional stakeholders include the IT team, employees who will use the upgraded infrastructure, and vendors who may need to provide technical support.

- Cybersecurity Protocol Implementation: You're a cybersecurity analyst tasked with implementing new security protocols across the organization. The primary intercept point on your Discover-Y mind map is the head of the IT department, who can facilitate or hinder the smooth implementation of these protocols. Additional stakeholders include all employees who will need to adhere to the new protocols, the legal team that should ensure compliance with regulations, and the executive board that needs to understand and support the importance of cybersecurity.
- Data Analysis Report: You're a data analyst preparing a report that could influence future business strategies. The primary intercept point on your Discover-Y mind map is the Chief Strategy Officer, who will use your insights to shape the company's strategic direction. Additional stakeholders include department heads who will implement the strategies, the finance team that will budget for any changes, and the data team that will provide you with necessary data.

Once you've chosen an encounter, write down your MAP for this situation. Remember, MAP stands for Meaningful (why it matters to you), Aligned (how it aligns with your values or goals), and Purposeful (what specific purpose it serves).

Next, outline your PAGE. Define the Purpose of the encounter, draft an Agenda, identify the Goal, and consider who you need Endorsement from for the Purpose, Agenda, and Goal.

Following this, draw out your Discover-Y mind map. Start with you at the base of the Y, your audience at the intersection, and the living and non-living elements associated with your audience on the two prongs of the Y.

Categorize questions between these components to deepen your understanding and uncover insights. Assign legends to the living category, such as trust, for a multidimensional approach.

Sample Questions:

- What is the meaningful aspect of this encounter for me?
- How does this encounter align with my professional or personal goals?
- What is the specific purpose of this encounter?
- Who do I need an endorsement from for my purpose, agenda, and goal?
- What are the living and non-living elements associated with my audience?
- Who does my audience trust? Who trusts my audience?

Allocate at least 1 hour 45 mins for this activity. Spend 15 minutes choosing and understanding the encounter, 20 minutes on MAP, 25 minutes on PAGE, and 45 minutes on Discover-Y. Remember, these are just guidelines. Feel free to spend more time where necessary.

By the end of this activity, you should have a broad understanding of how MAP, PAGE, and Discover-Y interact and complement each other. You'll be equipped with a practical approach to tackling workplace encounters effectively and efficiently. The aim is not just to complete the activity but to absorb these concepts so they can become second nature in your professional journey. I highly encourage you to run these activities with your mentor, manager, peer, or anyone else you trust. Iteration is key and use it to your advantage.

Elevate Your Engagement to Achieve Results

Imagine, if you will, walking into a challenging situation at work. Surprises abound, and the stakes are high. Fear not, for there is a tool, a subroutine if you will, that can help you navigate these treacherous waters with grace and acumen. It's as simple as remembering a child's song—Head, Shoulders, Knees, and Toes.

Let's start with the head, the seat of observation. Use your eyes—really see what is happening around you. Observe the dynamics, body language, and unspoken messages. This is the first step in understanding the problem.

Next, we move to the shoulders, where we often carry our burdens. Ask probing questions and delve beneath the surface to uncover the real problems. These issues weigh heavily on our shoulders, the crux of the matter.

Then we come to the knees, a symbol of supplication. Here, we need to unearth the real needs of the situation objectively. Strip away the superficial, and get down to what is truly needed. It's like bending the knee, acknowledging the true nature of the problem.

Finally, we arrive at the toes. This represents movement, progression, and the journey to a solution. Once you've thoroughly observed, questioned, and understood the problem, it's time to derive an appropriate solution. This is the next step forward, moving toward resolution.

The effectiveness of the Head, Shoulders, Knees, and Toes subroutine lies in its methodical approach to problem-solving. It encourages a holistic view of a situation, ensuring you don't miss any crucial details. By observing first, you gather data without prejudice. Investigating allows you to understand the root cause of the problem, avoiding knee-jerk reactions based on symptoms rather than causes. Identifying real needs ensures that solutions are tailored and effective, not one-size-fits-all fixes that may not address the actual issue. Finally, moving toward a solution encourages forward-thinking and progress.

First, let's consider a software developer who was tasked with creating a feature for an application. As described by the client, the feature seemed convoluted and unnecessarily complex. Rather than diving in and coding, the developer remembered the subroutine. He observed (head) that the client seemed frustrated when explaining the feature. He investigated (shoulders), asking probing questions about why this feature was needed. This led to the realization that the client was trying to solve a different problem (knees). Together, they derived a simpler, more effective solution (toes). The developer saved time, and the client was happier with the result.

Also consider an IT manager facing frequent network outages. Initial attempts to boost server capacity (the obvious solution) failed. Remembering the subroutine, she took a step back to observe (head) the patterns of the outages. She investigated (shoulders) by asking the team about their experiences during these outages. The outages happened at peak times when a particular team ran large data reports. The real need (knees) wasn't more server capacity but the staggered scheduling of large data tasks. They moved toward this solution (toes), effectively eliminating the outages without unnecessary expense on server capacity.

These anecdotes show that no matter your role in the tech industry, the Head, Shoulders, Knees, and Toes tool can help you navigate challenges more effectively. It encourages a deeper understanding of problems and fosters innovative solutions. This tool isn't about quick fixes but sustainable, effective problem-solving. Remember, the key is to know your customers, colleagues, and team better than they know themselves. This tool empowers you to do just that. When you crack this code, you open up a world of win-win outcomes. You'll often find that either no solution is needed or a low-cost, high-value alternative can be agreed upon.

Convince in Two Minutes or Less

Imagine you have to deliver an important message, a request to make, and just two minutes to do it. Sounds like a challenge, doesn't it? But here's a tool that can help—the two-minute message, or TMM for short.

The TMM is like a finely crafted script designed to be delivered in two minutes or less. It's perfect when you need to convince someone to see things your way. It works wonders in verbal communication, but it's equally effective in emails.

Now, this isn't some haphazardly thrown-together spiel. No, it's broken down into four key sections—audience context statements, key statement, supporting statements, and closing statements.

First off, you set the stage with your audience context statements. This isn't just about giving them the background information they need. It's about making them feel safe and comfortable. You're not there to fight, and you're there to engage. For instance, let's say you're asking for a promotion. You could start by acknowledging that your manager can agree or disagree with your request. This instantly creates a safe environment for discussion.

Next up is the key statement. This is your ask, plain and simple—no frills, no stipulations, just a clear, concise request that stands the test of time.

Then come the supporting statements. These are your reasons and justifications. Don't just rehash the past. Focus on the future and on the benefits your request could bring. This, my friend, is how you create leverage. It's like a promising athlete securing a lucrative contract based on their potential to bring success to the team.

Finally, you wrap things up with your closing statements. Instead of summarizing what you've already said, you seek endorsement from your audience. You let them know that you're available to support them in any way they need.

The beauty of the TMM lies in its brevity and clarity. It respects the recipient's time, gets straight to the point, and presents a compelling case in just two minutes. Let's consider a couple of scenarios.

A tech consultant has been trying to convince a client to adopt a new software solution. He used the TMM, first setting the context by acknowledging the client's concerns about change. His key statement was clear—the new software will streamline their processes. He supported this with statements about future efficiency gains and cost savings backed by examples from past clients. He closed by seeking the client's endorsement and offering his continued support.

A project manager needed additional resources for a critical project. She used the TMM to present her case to the department head. She set the context by acknowledging the budget constraints. Her key statement was simple—the project's success hinged on additional resources. She supported this with projections of future benefits regarding project outcomes and potential business growth. She closed by seeking the department head's endorsement and offering further details if needed.

The TMM is a powerful tool that can help you articulate your requests efficiently and effectively. It's not just about getting what you want. It's about fostering understanding, collaboration, and progress.

ACTIVITY: Bridging the Gap – From Observation to Convincing

This activity is designed to help you apply the concepts of Head, Shoulders, Knees, and Toes and the TMM in real-world scenarios. By working through this activity, you will learn how to navigate challenging situations at work and convincingly communicate your requests to stakeholders.

Materials Required:

- Your notebook for *Hurry Up Slowly*
- A quiet space
- An open mind ready for introspection

Instructions:

1. Choose one of the five workplace encounters listed below.
2. Using the Head, Shoulders, Knees, and Toes tool, write down your thought process as you navigate the situation. Structure your thoughts as follows:
   a. Head (Observe): Note what you see in the situation.
   b. Shoulders (Investigate): Write down the questions you would ask to understand the problem better.
   c. Knees (Identify): Based on your investigation, identify the real needs in the situation.
   d. Toes (Solve): Propose a solution that addresses the identified needs.
3. Once you work through this process, formulate your TMM. Remember to include:
   a. Audience Context Statements: Set the stage for your message.
   b. Key Statement: Clearly state your request.
   c. Supporting Statements: Provide reasons why your request should be granted.
   d. Closing Statements: Summarize your message and seek endorsement.
4. Write down your TMM in your notebook.

Sample Workplace Encounters:

- You are a software developer who believes switching to a new software development tool could improve team productivity.
- As an IT manager, you need additional budget for upgrading server infrastructure to avoid potential outages.
- You are a project manager who needs more team members for a critical project.
- As a data analyst, you want to implement a new data management system that could streamline operations.
- You are a cybersecurity specialist proposing a company-wide training session on phishing attacks.

Sample Questions and Their Intent:

- What are the current pain points with the existing system/tool/process? (This question aims to understand the problem)
- What are the benefits of implementing this solution? (This question is to unearth the needs and potential gains)

Allocate about 60 minutes for this activity. Spend around 20 minutes on the Head, Shoulders, Knees, and Toes process and 40 minutes crafting your TMM.

By the end of this activity, you should clearly understand how to navigate complex workplace situations using the Head, Shoulders, Knees, and Toes tool and how to communicate a request using the TMM convincingly. You will also have a practical example in your notebook that you can refer back to or adapt for future use. I highly encourage you to run these activities with your mentor, manager, peer, or anyone else you trust. Iteration is key, and use it to your advantage.

In the symphony of life, each component—the Head, Shoulders, Knees, and Toes—plays a vital role, akin to the different sections of an orchestra. The Head symbolizes our thoughts and strategies. Shoulders represent our strength to carry burdens. Knees signify the flexibility to adapt, and Toes keep us grounded in humility. Together, they orchestrate a balanced life.

On the other hand, the TMM is a reminder of the power of brevity, much like a poignant melody in a song. It emphasizes the importance of concise, impactful communication in this fast-paced world. So, my friends, remember to play each part of your life's symphony with gusto, and ensure every note of your communication resonates with clarity and purpose. Let these tenets be your guide on the path to success.

L.O.V.E.L.A.P.S: A Guide to Career Advancement Through Persuasion

Now, we dive into the fascinating world of persuasion and its role in your career growth. We will explore eight key concepts, each with a real-world example, and we'll tie them all together with a memorable acronym. So, let's get started!

Our acronym is L.O.V.E.L.A.P.S.

L - Loss Aversion (Lose to Gain)

The fear of losing something is often more powerful than the prospect of gaining something of equal value. Just look at insurance—we're willing to pay regular premiums to avoid the potential loss. In the workplace, this could translate to demonstrating the potential loss if your proposed solution isn't implemented. For example, an IT manager could highlight the potential loss of data or revenue if specific cybersecurity measures aren't implemented. Just like insurance companies convince us to pay premiums to avoid potential losses, you can use this principle to showcase the potential loss if your suggestions are overlooked. In 2007, Dropbox[23] showed users what they'd lose (i.e., easy file syncing and sharing) by not adopting their product, leading to a significant increase in sign-ups.

O - Opportunity Cost (Opt Out Cost)

Every time we choose to do something, we're also choosing not to do something else. Highlighting the cost of not acting on your proposal can be a persuasive tool. For instance, a project manager could diplomatically push back on an unfavorable idea by outlining what other opportunities the team would miss out on. Every decision has an opportunity cost. Highlighting what your team or company stands to lose by not acting on your proposal can be a persuasive tool. Netflix[24] convinced many cable users to cut the cord by highlighting the high cost of not switching—missing out on unlimited, on-demand content at a lower price.

V - Variety Thrills

People derive satisfaction from new experiences. As a tech professional, you could leverage this by constantly learning new skills and technologies, making you a valuable asset to any team. New experiences bring satisfaction. Constantly learn new skills to stay valuable and exciting to employers. Sir Richard Branson, the founder of the Virgin Group,[25] embodies the thrill of variety. Starting with a record shop, he expanded into airlines, telecommunications, space travel, and more, constantly seeking novel experiences and challenges. His diverse ventures demonstrate the benefits and excitement that can come from not being limited to one industry or skillset.

E - Economy of Choice

Offering too many choices can overwhelm and deter people. If you're a software developer proposing solutions, remember that three or four well-thought-out options will likely be more effective than a dozen alternatives. Too many choices can overwhelm. Limit options to boost engagement and decision-making. Steve Jobs simplified Apple's[26] product line to just a few key items, making the buying process easier for consumers.

L - Light and Dark Contrast

Our perception of things is affected by what they are not. As an IT professional, you might use this principle to demonstrate the effectiveness of a new tool or system by contrasting it with the old one. For example, "With our current system, we're blind. But with the proposed upgrade, it's like springtime." The contrast principle can make your proposals stand out. Show the stark difference between the current state and the potential future state. Slack[27] positioned itself as the remedy to chaotic, email-based communication, showing a clear contrast that attracted many teams.

A - Aspiring Potential

People are often more invested in what you could be than what you currently are. A rookie with growth potential might be valued more than a veteran whose growth has plateaued. When seeking a promotion, highlight your potential for future performance, not just your current accomplishments. Highlight your growth potential when seeking promotions or new roles. Potential often trumps past performance. Sundar Pichai's[28] potential was recognized by Google, leading him from product management to CEO.

P - Past is Past (Sunk Cost)

Decisions should be based on future gains, not past costs. If you've been working on a project proving unfruitful, don't let the time and resources already invested hold you back from pivoting to a more promising endeavor. Don't let past investments cloud your judgment. Future gains should drive decisions. IBM[29] divested its PC business to focus on more profitable ventures, ignoring the sunk cost in the PC division.

S - Set the Stage (Anchoring)

First impressions are powerful. They set the stage for all subsequent interactions. In salary negotiations, the first number offered often serves as an anchor that influences the rest of the negotiation. So, if you're negotiating your salary, aim high. Even if you have to come down, you'll likely end up higher than if you'd started low. The first impression sets the tone for all subsequent interactions. In negotiations, use anchoring to your advantage. Jeff Bezos[30] anchored Amazon as "Earth's biggest bookstore" before expanding into other product lines.

L.O.V.E.L.A.P.S isn't just a mnemonic device. It guides you in navigating your career, advocating for yourself, and driving your professional growth. Use it and watch as doors open and opportunities arise!

ACTIVITY: Harnessing L.O.V.E.L.A.P.S in Your Workplace

In this activity, you'll apply the principles of L.O.V.E.L.A.P.S to real-life workplace situations. You'll choose from five examples and write down your thought process for leveraging these principles to improve your sales and persuasion skills. This exercise will help you promote yourself and your objectives more effectively at work.

Materials Required:

- The notebook you've been using for this book
- A quiet place to think and reflect
- An open mind ready to apply new concepts

Instructions:

- Review the five example workplace encounters listed below.
- Choose one that you find most relevant or challenging. Feel free to customize the example to suit your situation better.
- For your chosen scenario, write down how you might apply each principle of L.O.V.E.L.A.P.S to improve your approach. Be specific and detailed in your responses.
- Reflect on your current approach to similar situations. How does it compare to the strategies you outlined using L.O.V.E.L.A.P.S?
- Finally, write down actionable steps based on your reflections. What will you do differently in the future?

Example Workplace Encounters:

1. Pitching a new project: You have an innovative idea for a project that could significantly benefit your company, but you need buy-in from senior management.
2. Negotiating a raise: You believe you're due for a pay increase, but you need to convince your boss.
3. Managing team conflict: Your team is divided over a decision, and you need to bring everyone on board with your solution.

4. Leading a product launch: You're in charge of launching a new product and need to convince stakeholders of its potential success.
5. Changing company processes: You believe adopting a new software or process could streamline work, but others in your company resist change.

This activity is self-paced, but we recommend taking one to two hours to complete. This will give you ample time to reflect on each principle and develop a comprehensive strategy.

By the end of this activity, you should have a clear, actionable plan for applying the L.O.V.E.L.A.P.S principles in your chosen scenario. You'll gain a deeper understanding of how these concepts can be used to improve your interpersonal effectiveness and career success. Remember, practice makes perfect! Try applying these strategies in your day-to-day interactions, and observe the results.

In the dance of life and business, we must all learn the steps to navigate the floor with grace and charm. L.O.V.E.L.A.P.S is your guide to this dance. Listen, Observe, Value, Empathize, Learn, Adapt, Persuade, and finally, Savor the sweet taste of success. Each step calls for an open heart, a keen eye, and a willing spirit. By truly listening to others, observing their needs, valuing their perspectives, and empathizing with their struggles, we invite connection and understanding. As we learn and adapt, we grow not just in knowledge but in wisdom. When we master the art of persuasion, we can influence others toward shared goals, creating a symphony of success. So let us dance, my friends, with L.O.V.E.L.A.P.S as our rhythm, and may each step bring us closer to our dreams.

SECTION 3: STORYTELLING

As you turn the page to this final section of Chapter 3, picture yourself as the master artist of your own life. Your canvas is the professional world. First, we'll delve into the art of marketing your brand—a skill as vital today as the air you breathe. You'll learn to present yourself like a confident artist unveiling a masterpiece through a captivating three-minute professional pitch. Remember, value is in the eye of the beholder, so always lead with outcomes and never undervalue your worth.

Next, we'll explore project planning with a twist—the project is you! Just as a sculptor carefully shapes the marble to reveal a beautiful statue, you'll learn to say no to what doesn't serve your career or the company. Use tools like the Goal Filter Tool and Knowledge Transfer Table to chisel away distractions and focus on your goals.

Finally, you'll master the art of writing and presenting your promotion letter. Think of yourself as a lawyer preparing a compelling case, using metrics to highlight your achievements. Always remember that your success is also your manager's success, so craft your case in a way that helps them champion your cause.

So, dear reader, let's embark on this exciting journey of self-improvement and career advancement together. This isn't just about securing a promotion; it's about sculpting the best version of yourself, both personally and professionally.

Market Your Brand

Craft Your Three-Minute Professional Pitch

As you embark on this journey of marketing your brand, the first step is crafting your three-minute professional pitch. The "Tell me about yourself" question may appear simple, but it's often overlooked or undervalued. It's like a well-traveled road that many have taken, but few have truly explored. So, let's take a different route, guided by psychology, business, and brevity principles.

Imagine your pitch as an art form. The brush strokes are your answers to eight key questions, each painted with no more than 40 words. The sequence in which you present them is crucial. First, you'll share four responses and then pause to gauge if the listener wants you to continue. This respects their time and creates a dramatic pause, enhancing the impact of your narrative.

Remember, this is not just a pitch. It's a conversation. Keep refining your answers, practice delivering them, and time yourself. With each iteration, your voice grows clearer, stronger, and more impactful. So, let's get started on this exciting journey to mastering the art of the three-minute pitch. Your future self will thank you for it.

Question 1: Why are you excited to be in the field you are in?

This question is about sharing your passion and establishing a common denominator with your listener. Remember, people are drawn to your "why" more than "what." Your excitement for your field should resonate with a greater cause that fuels your life purpose.

Question 2: What are the business outcomes you are responsible for?

Question 3: What are the organizational goals you are responsible for?

Question 4: What are the daily-weekly deliverables you are responsible for?

These questions are crucial because they shift the focus from activities to results. Rather than discuss the busywork that consumes your day to day, you highlight the tangible outcomes that your work produces.

Leading with outcomes is a powerful strategy because it directly communicates the value you bring to an organization. It's not about the volume of tasks you handle but the impact of those tasks on the company's bottom line.

When you discuss the business outcomes, organizational goals, and deliverables you're responsible for, you paint a clear picture of the results you can deliver. This allows potential employers to immediately understand your worth and envision how you fit into their operations.

Moreover, focusing on outcomes rather than busywork helps you distinguish between what's merely urgent and what's truly important. It encourages you to prioritize tasks that contribute to key objectives and let go of tasks that do not.

Therefore, when crafting your three-minute pitch, remember to lead with outcomes. Discuss the results you achieved, the goals you reached, and the deliverables you produced. This approach will make your pitch more compelling and demonstrate your focus on results, which is what most employers look for.

At this juncture, take a pause. Ask if they're interested in hearing more. If not, remember, it's not a reflection of your worth. Many might not see or want the value you offer, and that's okay.

Question 5: In what capacity do you operate?

Here, share the various roles you undertake despite your official title. Use any hardships or unfairness as an opportunity to showcase your resilience and versatility.

Question 6: What are the self-development goals you have for yourself?

Share how you're committed to growth, not just for job requirements but also for personal development. Your commitment to reading this book already testifies to your dedication toward self-improvement.

Question 7: What makes you unique?

Dig deep for this one. Avoid generic statements and find what truly sets you apart. If you struggle, seek feedback from trusted individuals.

Question 8: How do you help others, and what are your thoughts on scaling it?

Finally, discuss how you contribute to the larger community. This is especially important for those aspiring to or already in people-management roles.

Remember, this pitch is a conversation, not a monologue. Keep refining your answers and practice delivering them. Each iteration brings you closer to a pitch that truly represents you and your value.

The three-minute pitch from the lens of a senior manager or director who has seen countless pitches and interviewed a multitude of candidates:

> *Firstly, understand that from our perspective, every minute is valuable. When you pitch, you're not just selling your skills or experiences. You're selling us time. We're investing our precious minutes into listening to you. So, make it count.*
>
> *Secondly, we're not just looking for someone who can do the job. We're looking for someone who fits into our culture, shares our values, and can contribute to our vision. Your excitement about your field (Question 1) gives us an idea of whether you're merely looking for a job or are genuinely passionate about what you do.*
>
> *When you talk about your value to your current organization (Questions 2, 3, and 4), we're not just hearing about your accom-*

plishments. We're also gauging how you perceive your role and its impact on the larger business objectives.

The capacity in which you operate (Question 5) and your self-development goals (Question 6) give us an insight into your potential for growth. It tells us whether you're comfortable with the status quo or are constantly striving to improve.

What makes you unique (Question 7) isn't just about what sets you apart from other candidates. It's about how you could bring a fresh perspective or new ideas to our team.

Lastly, how you help others and your thoughts on scaling it (Question 8) reflect your leadership potential. In the tech world, technical skills are essential, as is the ability to work as a team and help others grow.

Remember, we're not just hiring a set of skills. We're hiring a person. Your three-minute pitch is your opportunity to show us who you are, not just what you can do. Make it count.

ACTIVITY: Crafting Your Three-Minute Pitch

This activity aims to help you create a compelling three-minute pitch that effectively communicates your professional identity and value. By the end of this activity, you should have a draft of your pitch ready for practice and refinement.

Materials Required:

- The notebook you've been using for this book.
- A quiet space where you can think and write without distractions.
- A timer (optional).

Instructions:

1. **Identify Your Why:** Reflect on your excitement about your field. Consider personal stories or experiences that led you to choose your profession. Write down your thoughts in your notebook.
2. **Define Your Value:** Consider the business outcomes, organizational goals, and daily deliverables you're responsible for in your current role. Be specific and use concrete examples. Write these down.
3. **Recognize Your Capacity:** Identify your role within your team or organization. Are you a manager, a team lead, or an individual contributor? Note down the scope of your responsibilities.
4. **Outline Your Self-Development Goals:** Reflect on your aspirations for personal growth. What skills are you currently developing? What additional responsibilities do you hope to take on in the future?
5. **Determine Your Uniqueness:** What sets you apart from others in your field? This could be a unique skill, an unusual experience, or a novel approach to problem-solving. Write this down.
6. **Describe How You Help Others:** Explain how you assist others in your work. This could mean mentoring junior colleagues, volunteering in your community, or implementing policies that improve the workplace for everyone.
7. **Craft Your Pitch:** Now, combine all the elements above into a cohesive three-minute pitch. Answer the eight questions described earlier.

Set aside at least an hour for this activity. However, crafting a good pitch takes time and shouldn't be rushed. You might need to revisit and refine your pitch over several days or weeks.

By the end of this activity, you should have a draft of your three-minute pitch ready. It should be concise, engaging, and representative of your professional journey and aspirations.

Remember, your pitch isn't set in stone. As you evolve in your career, your pitch should evolve too. Regularly update and practice your pitch, refining it as necessary.

Project Plan the Ultimate Product—You!

In the words of the great Jim Rohn, "Either you run the day, or the day runs you."[31] To truly take control of your career and life, you need to become the master of your time and the architect of your priorities. Here, I introduce a powerful tool to help you do just that.

The Goal Filter Tool

The Goal Filter Tool is like a compass for your professional journey. It helps you navigate the vast sea of tasks and deliverables to reach the shores of your most important goals.

This simple yet effective tool works on three vectors to help you assess every task or deliverable that comes your way:

1. Business Importance: How vital is this task to the business? Is it a cornerstone for your company's success or merely an accessory?
2. Urgency: Does it require immediate attention, or can it wait?
3. Career Impact: Will completing this task contribute significantly to your professional growth and advancement?

Here's the key: A task must score a 1 on all three vectors for immediate attention. But beware, the threshold for a 1 is extremely high. It should be in the 90th percentile. Why so stringent, you may ask? Because if everything seems important, then nothing truly is.

Setting such a high bar ensures that only the most critical and impactful tasks receive our immediate attention. Everything else can be delayed, delegated, or dropped.

The beauty of the Goal Filter Tool lies in its objective clarity. It empowers you to make informed decisions about where to invest your time and energy. It provides tangible proof to have objective dialogues with your managers and stakeholders about your priorities and responsibilities.

As Jim Rohn said, "Don't mistake movement for achievement."[32] The Goal Filter Tool ensures that every step you take is toward meaningful progress, not just busywork. It's a tool that turns you into the project manager of the most focal product—You!

So, I urge you to apply the Goal Filter Tool professionally. Experience the clarity it brings, appreciate its power, and share its wisdom with others. In the end, it's not about how busy you are but how much you accomplish that truly matters.

ACTIVITY: Applying the Goal Filter Tool and Discussing It with Your Manager

This activity will guide you through using the Goal Filter Tool to prioritize your tasks and deliverables and advise you on how to discuss this with your manager effectively.

Materials Required:

- Your notebook
- A pen

Instructions:

1. List Your Tasks: List all your current tasks and deliverables in your notebook.
2. Score Each Task: Next, score each task on three vectors—business importance, urgency, and career impact. Remember, the threshold for scoring a 1 is extremely high, i.e., it should be in the 90th percentile.
   a. Business Importance: If the task is vital to the business and its success significantly depends on it, score it as 1. Otherwise, give it a 0.
   b. Urgency: If the task requires immediate attention and delaying it might have serious consequences, score it as 1. Otherwise, give it a 0.
   c. Career Impact: If the task will contribute significantly to your professional growth and advancement, score it as 1. Otherwise, give it a 0.
3. Apply the Filter: If a task scores 1 in all three vectors, it's a task that you need to focus on right away. Tasks that score less than 1 on any vector are tasks that you may consider to delay, delegate, or drop.
4. Prepare for the Discussion: Understand your reasoning behind each score. This clarity will help you articulate your points effectively with your manager.
5. Request a Meeting: Request a dedicated meeting with your manager to discuss your tasks and deliverables. This shows respect for their time and allows them to focus solely on the conversation.
6. Communicate Effectively: In the meeting, explain the Goal Filter Tool and how you used it to score your tasks. Lead the conversation with outcomes and how focusing on high scoring tasks can lead to better results.

7. Show Empathy and Be Open to Feedback: Understand your manager's perspective and be prepared to listen and accept feedback. Your manager may have additional information that could affect the scoring of your tasks.
8. Propose Solutions: Suggest ways tasks that score low on the Goal Filter Tool could be delegated, delayed, or dropped without impacting overall productivity.
9. Follow-up: After the meeting, send a follow-up email summarizing your discussion and any agreed-upon next steps.

This activity should take about 1-2 hours. However, the time will vary depending on your number of tasks and the length of your discussion with your manager.

By the end of this activity, you should know which tasks are most important and urgent and which tasks contribute the most to your career growth. This will allow you to focus your efforts where they matter the most, leading to greater productivity and professional advancement.

Remember, the Goal Filter Tool is not a one-time activity. It's a continuous process of assessment and reassessment as your tasks and priorities evolve. So, make it a habit to regularly apply the Goal Filter Tool and keep your focus where it truly belongs.

Introducing the Knowledge Transfer Table: A Tool for Leadership in Absence

As we navigate the journey of our careers, we encounter numerous situations that test our resolve, resilience, and leadership skills. We all have to grapple with the reality of poor leadership at times, but remember, managers are humans too, with their strengths and weaknesses. In such situations, it is up to us to step up and fill the void.

Let me introduce you to a powerful tool that can help in this regard—the Knowledge Transfer Table.

This tool is particularly useful in two specific scenarios: when you're going on an extended leave, like a sabbatical or bonding leave, and when you're leaving your current team, either for another company or another team within the same company. In both cases, using the Knowledge Transfer Table can minimize impact, set clear boundaries, and prevent feelings of remorse or abandonment.

The Knowledge Transfer Table is a simple yet powerful tool that lists all your tasks and deliverables, assigns priorities, identifies current and expected progress, recommends delegates, provides brief descriptions and key points, and even includes links to audio or video files for detailed explanations. This serves as your "hologram," ensuring seamless continuity of work during your absence.

Using this tool has several psychological benefits. For the delegate, it provides clarity and direction, reducing anxiety and confusion. For the manager, it reassures them that work will continue smoothly, relieving them of additional stress. And for you, it offers peace of mind knowing that you've done your part to ensure a smooth transition. This proactive approach toward responsibility and accountability truly marks the birth of a leader.

By documenting and communicating your work in this way, you demonstrate leadership in its truest sense. It is a sign of maturity, foresight, and empathy. It acknowledges the human element in the workplace, understanding that change can be challenging but also managed effectively with the right tools and approach.

Methodical Guide to Using the Knowledge Transfer Table

The Knowledge Transfer Table is a structured and efficient tool designed to ensure seamless transitions during periods of absence.

Here's a step-by-step guide to help you understand, relate, appreciate, and apply this tool effectively.

STEP 1: Elements in Table Form

Begin by setting up the table with the following columns: Tasks, Priority, Current Status (%), Expected Handoff (%), Return (%), Delegate, Description and Key Points, and Audio/Video Links.

TASK	PRIORITY	CURRENT STATUS %	HANDOFF %	RETURN %	DELEGATE	DETAILS	AUDIO/ VIDEO LINKS
Item 1	1	10%	35%	100%	Person A	Check with Mkt	link1
Item 2	1	65%	85%	100%	Person B	Notes...	link2
Item 3	0	70%	75%	85%	Person C	----	link3

STEP 2: List Tasks and Deliverables

In the Tasks column, list all your ongoing tasks and deliverables. These are the activities you're responsible for and will need someone else to manage in your absence.

STEP 3: Assign Priorities

Assign a priority to each task based on its importance and urgency, with 1 being high priority and 0 being low priority. This will aid your delegate in understanding which tasks need immediate attention.

STEP 4: Identify Current Status

In the Current Status (%) column, note down the percentage of each task that has been completed. This gives your delegate a clear picture of how much work has already been done.

STEP 5: Determine Expected Handoff and Return Percentage

Next, determine the Expected Handoff (%), which is the level of completion you expect to reach before handing off the task. Then, estimate the Return (%) or the completion status you expect the task to reach when you return. This helps set expectations for progress during your absence.

STEP 6: Recommend a Delegate

Identify who you recommend to take over your tasks in the Delegate column. If possible, ask this person to secure their agreement and the manager's approval. This ensures they are not taken by surprise and can be prepared accordingly. Remember to distribute the tasks evenly among delegates to avoid overloading anyone.

STEP 7: Provide a Brief Description and Key Points

In the Description and Key Points column, explain each task concisely. This helps your delegate understand the nature of the task and any critical points related to the project.

STEP 8: Record Audio/Video Files

Finally, in the Audio/Video Links column, attach links to audio or video files where you explain each task in greater detail. Recording these explanations can be more efficient than writing everything down and allows you to communicate more information.

By following these steps methodically, you can effectively use the Knowledge Transfer Table to manage your transitions, ensure continuity of work, and minimize disruptions. Not only will this tool help maintain productivity, it will also foster a culture of accountability and leadership within your team.

The Knowledge Transfer Table is more than just a tool for handing over tasks. It is a testament to your commitment to your team, role, and organization. It shows that you are a leader who cares about continuity, productivity, and the well-being of your team. In doing so, it helps you manage transitions smoothly and leaves a lasting positive impact on those you work with.

As you progress in your career, remember that great leaders aren't just born; they are made through consistent effort, resilience, and a deep commitment to nurturing their teams, even in their absence. Embrace the opportunity to step up when leadership is missing and become the leader you wish to see in your workplace. The Knowledge Transfer Table is your ally in this journey. Use it wisely and watch as it paves the way for overall success.

Embrace the power within you, my friends! The Goal Filter Tool and Knowledge Transfer Table are more than mere instruments. They're your secret allies in the quest for professional excellence.

Picture this: a mountain of objectives before you, and the Goal Filter Tool is your compass, guiding your tasks to align with your ultimate vision. Each ticked-off task isn't mere work done. It's a stride toward your grand ambition.

When life calls you away from your desk, fear not! The Knowledge Transfer Table steps in, ensuring that your teammates are not left in disarray but are empowered to carry your work forward with ease.

So, let us weave these tools into our work fabric, fostering transparency, accountability, and leadership. For these aren't just tools;

they're your bridges to an exciting future, your testament of resilience, and your stepping stone to creating win-win situations. It's time to raise the bar and redefine success. The future is thrilling, and it starts today!

Writing and Presenting

> *"I am smart. I am kind. I am rich. I am abundant. I am strong. I am self-sustaining. I am better. I am victorious. I am the growth. I am character. I am increased. I am lightning. I am undeniable. I am relentless. I am incredible. I am the universe. I am the savior. I am positive. I am the beast. I am the reward. I am the answer. I am the separation. I am the result. I am bigger. I am here. I am now. I am me. I am what I'm meant to be. I am a work in progress. I am progress. I am the wisdom. I am the vision. I am what the world needs. I am change. I am power. I am gentle. I am past, present and the future. I am You. And You are the 'I' in this message."*

Embrace the power of your emotions, my friends! They make us human, the very lifeblood of our writing and presenting. It's essential to understand that sharing our emotions is a delicate dance—for they can be used against us, reflecting not our flaws but the insecurities and imperfections of others. Yet don't let this paint a picture of doom and gloom. We hold the reins of our emotions, and it's within our grasp to channel them, not just for our benefit but for the enrichment of others.

Our journey so far has taught us that people are drawn not to what we do or create but to why we do it—a gem of wisdom from Simon Sinek. So, let's extrapolate this philosophy to our writing and presentation. Instead of adorning our communication with fancy words or beautiful visuals, let's focus on the why, the emotion, and the intent behind our messages. Eloquence in English is not a measure of intelligence. True brilliance can speak Spanish, Mandarin, Tamil, or any other language, sparking innovation and change.

As we dive deeper into technical fields, we must be aware of the trap of knowledge bias and the danger of complexity. The key to excellent communication lies not in complexity but in simplicity. Can a fifth grader understand you? Then, you're on the right path. Our young friends can teach us much about directness and honesty if we set aside our egos and listen. This journey of self-improvement helps us catch our biases, blind spots, and oversights.

Now, consider the piece of emotional storytelling written three paragraphs ago: "I am smart. I am kind. I am rich..." Each phrase rings with affirmation; each sentence pulsates with emotional resonance. It's a testament to the power of storytelling, a compelling example of how emotive language can draw people in and inspire them. This is the power you wield when you harness your emotions in writing and presenting. So, let's step into this journey together, embracing our emotions and using them to shape our words and our world.

From Paper to Podium: How Solid Writing Fuels Unforgettable Presentations

Emotions are more than just feelings; they're your secret weapon in storytelling. Use them, embrace them, and watch your narrative catch fire. There are a couple of misconceptions about injecting emotion into your writing. One, it's unprofessional or overly sentimental. This is not the case. When used appropriately, emotion adds depth and relatability to your writing. It's about striking the right balance. Another misconception is that all emotions are suitable for all types of writing. The type and intensity of emotion should suit the context and purpose of your writing. A marketing professional writing a product description might tap into their excitement about the product's features, conveying this enthusiasm to potential customers. Don't just write words on a page; put your heart into it! Feel the thrill, the pain, the joy, and let it all spill onto that page. Your readers want to connect with a real person, not a robot.

What message are you dying to share? That's your guiding star—your anchor. Keep it front and center in your writing and never lose focus. When it comes to identifying your "why," some people think it's the same as your topic. But it's not. Your "why" is the underlying purpose or message of your writing, not just the surface-level subject. Another misconception is that your "why" can be vague or broad. In truth, a clear, specific "why" gives your writing focus and direction. A business executive preparing a company-wide memo might identify their "why" as wanting to inspire employees to embrace a new strategic direction. Ask yourself—why are you doing this? What's the endgame? Every word you write should serve that purpose. If it doesn't, cut it out. Your writing should reflect your mission, goals, and hustle.

If a kid can't understand your message, you're not doing it right. The most powerful ideas are simple. Strip away the fluff and get to the heart of it. Simplicity in writing also has its misconceptions. Some people believe simple writing is simplistic or childish. But simplicity in writing means clarity and accessibility, not a lack of sophistication or depth. There is also a belief that complex language makes your writing sound more intelligent or professional. However, clear, straightforward language is more likely to be understood and appreciated by a wide range of readers. An HR manager creating an employee handbook might strive for clear, straightforward language to ensure all policies are easily understood. Don't hide behind fancy words or complicated sentences. Keep it simple. Keep it real. People don't have time to decipher your message. Tell it straight and make every word count.

Critiques aren't personal attacks; they're golden opportunities for growth. Welcome them, learn from them, and watch your writing evolve. Feedback often gets a bad rap. People tend to think feedback is always negative or critical. However, constructive feedback is a valuable tool for learning and growth, not an attack on your abilities. Another common misconception is that you have to implement all feedback you receive. On the contrary, you should consider all feedback, but use your judgement to decide what will truly improve your work. A project manager might seek feedback on a project plan from team members, incorporating their suggestions to improve the plan's effectiveness. You're not perfect, and neither is your writing. Listen to what others have to say. Take their feedback, learn from it, and grow from it. But at the end of the day, you make the final call.

Your voice, style, and truth are what readers connect with. Be unabashedly you, and watch your authenticity echo with your audience. Staying authentic in your writing can be misunderstood too. Some people believe authenticity means sharing all aspects of yourself or your feelings. But authenticity in writing is about maintaining your unique voice and perspective, not oversharing. Another misconception is that you need a dramatic or unusual life to be authentic in your writing. The truth is authenticity comes from being true to your experiences and perspectives, no matter how ordinary they might seem. An entrepreneur pitching to investors might stay true to their passion and vision for their company, rather than trying to fit what they think investors want to hear. Be unapologetically you. Don't try to sound like someone else. Your unique perspective is your strength—your secret weapon. Use it. Let your authenticity shine through every word you write.

Mastering these writing skills sets the foundation for effective presentation skills. When you know how to write with clarity, authenticity, and emotional resonance, you'll be able to translate those skills into engaging and impactful presentations. It's about taking the compelling narratives you've crafted in writing and bringing them to life verbally. It's about connecting with your audience, not just through words but also voice modulation, body language, and passion. So, hone your writing skills today, and watch as they elevate your presentations tomorrow!

Pause, Play, Slay: Unleashing the Confident Speaker Within

Now, let's take a journey into the realm of powerful presenting! We're going to unlock seven secrets that will transform you into a captivating speaker, whether you're talking to your classmates or, one day, to your colleagues in the tech world!

1. Outcome Oriented

First, always begin with the end in mind. Imagine you're building a robot. You wouldn't start without knowing what you want the robot to do, right? The same goes for your speech. Know your desired outcome and steer your words toward that goal. A common misconception while delivering a presentation is that an unwavering focus on the desired outcome may lead to rigidity, leaving little room for unexpected adjustments. This can be effectively managed by anticipating various scenarios and being flexible to modify your approach when necessary. For instance, in a scenario where you unveil a new software feature to your team, your primary goal could be to ensure their understanding and acceptance. However, if you encounter unforeseen queries or concerns, it's crucial to address them, even if it means straying slightly from your initial plan.

2. Clarity Quest

Next, make sure everyone's aboard your spaceship of thoughts. Ask questions like "Are you with me?" or "Does that make sense?" It's like checking if all systems are go before blasting off to the next part of your speech. Often, people believe that clarity in speech is synonymous with using simple language. This understanding is incomplete without considering the importance of structuring your thoughts and ideas logically. For example, when explaining a convoluted coding problem, dissect it into smaller, manageable parts and elucidate each clearly before proceeding to the next segment.

3. Assertive Adventure

Here, you need to stand tall and speak with confidence. It's like being the captain of your spaceship—you need to show you're in control. Speak clearly and firmly, but remember, a good captain respects their crew just as much as they command respect. Assertiveness is sometimes misconstrued as aggression or domineering behavior. It's crucial to dispel this myth by balancing assertiveness with empathy and respect for others' viewpoints. Suppose you disagree with a colleague's approach to a tech issue. In such cases, it's beneficial to confidently propose your alternative solution while acknowledging their perspective and encouraging a discussion.

4. Speed Spectrum

Now, we regulate our speech speed. If you talk too fast, it's like driving your spaceship at light speed—everyone gets lost. Too slow, and it's like floating aimlessly in space. Find a steady speed where everyone can enjoy the journey. There's a prevalent belief that rapid speech is indicative of intelligence and competence. However, clarity and effective communication often necessitate a moderate pace to allow listeners to thoroughly absorb the information. For instance, when demonstrating a new algorithm, instead of rushing through the details, take the time to explain each step comprehensively.

5. Volume Voyage

Here, we turn up the volume dial, not too loud that it hurts your crew's ears and not too soft that they can't hear your commands. Find the right volume that is clear, comfortable, and compelling. People often assume that speaking louder will automatically command attention. Contrarily, strategically varying your volume can highlight key points and maintain audience engagement. For example, suppose you're announcing a significant breakthrough in your tech project. In that case, you could elevate your volume for the announcement to underscore its significance.

6. Rhythm Route

Say goodbye to monotone and hello to rhythm. Imagine your words are like music. A monotone speech is like a song with only one note. Add different notes—highs and lows—to make your speech more interesting. Rhythm in speech is often overlooked, with many assuming it's more relevant to music or poetry. However, incorporating rhythm can make your speech more engaging and memorable. When explaining a technical process, try to vary the rhythm of your speech to avoid sounding monotonous and to underscore crucial points.

7. Pause Pitstop

Finally, we come to the power of pauses. It's like stopping at a space station to refuel and reflect. Pauses give your audience time to process your words. Decisions are often made in silence, so use pauses wisely. Pauses during a speech are commonly interpreted as signs of uncertainty or lack of preparation. However, strategic pauses can provide your audience with time to process information and also contribute to building suspense or emphasis. For instance, if you're about to reveal a new tech innovation, a pause before the revelation can heighten anticipation.

These are your seven secrets to becoming an exceptional presenter. Practice them, make them your own, and see how they transform your communication skills. Remember, great speakers aren't born—they're made. So, step into your spaceship and let's blast off to the stars of powerful presenting!

ACTIVITY: Boost Your Writing and Presentation Skills

Embrace the art of storytelling by mastering writing strategies that will not only enhance your written communication but also set the foundation for effective presentations. This hybrid activity combines powerful writing techniques with practical exercises to help you write with clarity, authenticity, and emotional resonance while also teaching you how to deliver engaging presentations.

Materials Required:

- Your notebook
- A quiet space to reflect and write
- An open mind ready to learn and apply new techniques
- A recording device (optional)

Instructions:

- **Feel It, Write It:** Reflect on a piece of writing where you injected an emotion. Evaluate its impact and note down how it could be improved. Emotions add depth and relatability to your writing and can also make your presentations more engaging.

- **Find Your Why:** Identify a message or purpose in your writing that drives you. Also, reflect on a past presentation or a future one you plan to give. Write down the desired outcome of this presentation. Your "why" gives your writing and your presentations focus and direction.
- **Keep It Real, Keep It Simple:** Write a complex idea in simple terms. Now try to break it down into its simplest form, making sure it's easy to understand. This exercise will ensure your audience grasps your message, both in your writing and in presentations.
- **Feedback Is Your Friend:** Seek constructive feedback on your writing from someone you trust. Also, record yourself giving a part of your presentation and listen back, noting down areas where you spoke too fast or too slow. Constructive feedback and self-assessment are valuable tools for learning and growth.
- **Authenticity Wins:** Write a paragraph about something you're passionate about. Make sure it reflects your unique voice and perspective. Also, think of a time when your opinion was different from others in a presentation. Write down ways you could improve your assertiveness without disrespecting others' ideas. Your unique voice and perspective should shine through in both your writing and your presentations.
- **Volume Voyage:** Listen to your recorded speech. Identify parts where your volume was too high or too low. Adjusting your volume strategically can emphasize key points and keep your audience engaged in your presentations.
- **Rhythm Route and Pause Pitstop:** Write down a part of your speech. Read it aloud and play with the rhythm. Try different rhythms until you find one that makes your speech more engaging and less monotonous. Also, identify places in your presentation where a pause might give your audience

time to process or build suspense. Incorporating strategic pauses can make your speech more impactful.

This activity should take about two to four hours. Take your time to reflect, write, and make meaningful changes.

At the end of this activity, you should have a clear understanding of how to write with clarity, authenticity, and emotional resonance and how to deliver powerful presentations. You'll be able to translate these skills into engaging and impactful presentations, connecting with your audience, not just through words but also voice modulation, body language, and passion. Your writing and your presentations should be a reflection of your authentic self, telling your unique story in a way that resonates with your audience.

CHAPTER SUMMARY

As we wrap up this chapter, let's reflect on the remarkable transformation you've undergone. You became a beacon of clear values in your professional life, shaping decisions that align with your core principles. Your newfound ability to say no has given you control over your time and energy, allowing you to focus on what truly matters. Your questioning skills sharpened, leading to deeper insights and stronger leadership. You built a solid network of individuals who share your vision and support your growth. Your skills in connecting with your audience and influencing others blossomed, enhancing your effectiveness in all interactions. With your understanding of sales and persuasion techniques, you're now a formidable negotiator, capable of driving action. Your personal brand shines brighter than ever, and you learned to focus on outcomes, propelling you beyond your perceived potential. Your project planning and personal development skills improved, leading to optimized goals and a more efficient workflow. Lastly, your writing and presentation skills were honed to perfection, enabling you to communicate more effectively and tell compelling stories. Carry these accomplishments with you as you continue on your journey to professional excellence. Become the leader you wish you had!

CHAPTER 4

Tidying Up Loose Ends: There Is More To Career Development

SECTION 1: CONSISTENCY – THE HIDDEN POWER YOU'RE IGNORING

It's not about those "eureka" moments or instant fame. It's the daily grind—the never-ending hustle toward your goals. However, lack of consistency is a silent killer, slowly and steadily sabotaging your dreams. If you don't get a grip on it, it can turn your big dreams into mere fantasies. Intrigued? Dive in to learn how unleashing the beast of consistency can flip your professional game.

Consistency isn't some fancy term coined by self-improvement evangelists. The unsung hero, the hidden gem, distinguishes the achievers from the dreamers. Consistency is about building momentum. It's the art of repeating an action until it becomes as automatic as breathing. It's not a conscious effort but an instinctive drive. This is where you witness genuine progress and growth. But it doesn't stop at personal development. Consistency is also about reliability—

about being someone others can count on. A consistent person is credible, consistently delivering on promises, making them a beacon amidst the crowd.

In the realm of business, consistency is the golden key. It's all about establishing recognition, fostering trust, and ensuring top-notch performance each time. Consistency doesn't make you a face in the crowd. It makes you the face of the crowd—the one they recall, the one they return to. Let's also remember the role of consistency in self-discipline and self-control. With consistency, you're the pilot of your life, not a passenger to your circumstances. You're the decision-maker, the limit-pusher, and the barrier-breaker. So, prepare yourself to harness the power of consistency. It might be a bumpy ride, but rest assured, it will be worth every twist and turn.

Real-Life Examples of Successful Entrepreneurs and Their Commitment to Consistency

Let's embark on this exploration. We will spotlight global dynamos who mastered the art of consistency.

First, we have Huda Kattan, the beauty queen helming Huda Beauty, a brand that hails from the sands of Dubai. This remarkable woman successfully channeled her passion for beauty into a billion-dollar empire.[33] But there's more to her story than just financial success. The cornerstone of Kattan's triumph lies in her unwavering commitment to consistency, particularly in community-building. She isn't just selling beauty products; she's cultivating a devoted tribe, a community that consistently engages with her brand. She leverages social media as an extension of her persona, consistently offering value, unveiling beauty secrets, and revealing her authentic self. Her dedication to consistency is evident in her regular engagement with her audience. Whether it's through her informative blog posts, interactive Instagram stories, or insightful YouTube tutorials,

Kattan never misses an opportunity to connect with her followers. This consistent interaction has not only helped her build a loyal fanbase but also allowed her to understand their needs better, leading to creating products they genuinely love. If you wish to mirror some of Kattan's strategies into your journey, pinpoint your passion and build a community around it. Consistently show up, offer value, and let authenticity guide your actions. Remember, in business and beyond, consistency is not just about repetition; it's about reinforcing trust, establishing credibility, and fostering a sense of belonging among your audience. And that, dear friend, is your ticket to success.

Moving on, we have Cristina Junqueira, the cofounder of Nubank,[34] a gamechanger in Brazil's fintech space. This visionary woman is not just challenging the traditional banking norms but is also creating a pathway for women in the tech industry. What makes Junqueira stand out? It's her unwavering commitment to consistency. Despite obstacles and setbacks, she relentlessly perseveres, consistently breaking barriers, taking calculated risks, and driving change. She embodies consistent courage, never shying away from challenges or norms that need challenging. But it's more than just her journey. Junqueira's commitment to fostering diversity and empowering women inspires aspiring female leaders in Brazil and beyond. Her dedication isn't limited to her craft alone but extends to the entire community she is building. If you're aiming to walk in Junqueira's footsteps, cultivate a relentless attitude. Be consistently brave, embrace risks, challenge norms, and persist no matter what. Take a leaf out of Junqueira's book and understand that true success lies not only in consistent efforts but also in consistent learning, growing, and evolving. Always remember the power of consistency is your greatest ally on this journey.

On another note, let's turn our attention to Richard Branson,[23] the genius behind Virgin Group. His secret is an unquenchable thirst for knowledge. He's continually evolving, learning, and growing, whether it's through reading a riveting book, immersing himself in enlightening workshops, or attending thought-provoking conferences. But it's not merely about accumulating information. It's about consistently pushing boundaries, challenging his perspectives, and expanding his horizons. Yet it's his commitment to consistency that truly sets him apart. Branson once said, "I have always lived my life by making lists. These vary from lists of people to call, lists of ideas, lists of companies to set up, lists of people who can make things happen."[23] This is a testament to his consistent approach to planning and execution. If you aspire to incorporate Branson's mindset into your journey, never stop questioning and exploring. Consistently feed your curiosity and strive to learn something new every day. And just like Branson, maintain a consistent approach to your actions, whether making lists or setting goals. The more you learn, the more you'll realize how much you don't know. That's where true wisdom resides.

So, whether you're building communities like Kattan, breaking barriers like Junqueira, or feeding your curiosity like Branson, the power of consistency is your greatest ally. It's about showing up, day after day, committed to your growth and success.

Let's take inspiration from the story of "The Hundred Penny Box" by Sharon Bell Mathis. This is a touching story about a boy named Michael and his great-great-aunt, who has a box of 100 pennies, each representing a year of her life.

Now, think about your life as that 100-penny box, with each penny representing a day of effort or change. If you go through your days without adding anything new or making any improvements, you'll have a box with 100 pennies—symbolizing a life lived but not necessarily grown.

But imagine if you decided to add a penny—a small, positive change—to your box every day. At first, it might not seem like much. After a week, you've added seven pennies. After a month, that's 30 extra pennies. After a year? You've added 365 pennies to your box!

Suddenly, your 100-penny box isn't just a box of 100 pennies anymore. It's a box with 465 pennies, each representing a day you chose to grow, change, and improve.

That's the power of small, consistent changes. They may not seem significant, but they can transform your 100-penny box over time into a treasure chest of growth and progress. So, what "penny" will you add to your box today?

Leveling Up in Life: How Video Games Taught Us About Perseverance

Imagine Samarth Hegde stepping into his summer internship with Grounow in 2023. With a fiery passion for video games and a thirst to improve his ScrumMaster and program management skills, Samarth was poised for an extraordinary journey.

We devised a project that was as unique as it was transformative. We merged Samarth's love for video games with a structured learning experience. The reward was more gaming time with his friends—a simple yet powerful motivator. This wasn't just about gaming. It was about turning a hobby into a tool for growth, nurturing leadership skills, instilling consistency, and fostering a strong work ethic.

So, we formed a scrum team, with Samarth leading the charge. His project became to play video games every day for an hour. This might seem trivial, but over a year, it added up to 172 additional hours of gaming—well-earned and well-spent time. But there was a twist. Samarth and his friends had to complete real-life tasks to earn their gaming time. They sent gratitude notes, performed chores, and finished homework. These were simple actions but ones that taught them responsibility, gratitude, and the pleasure of earning their rewards. Physical activity was an integral part of the project, too. 100 jumping jacks before and after each gaming session. This kept them active and reminded them of the importance of balance between leisure and health.

After each gaming session, they sat together, discussing their experience, reflecting on their strategies, and planning for the next session. This instilled in them the art of self-reflection, the importance of feedback, and the value of continuous improvement.

Anticipating objections and responding with reasonable commitments were crucial to their journey. This taught them the power of negotiation, the importance of setting boundaries, and the art of achieving win-win outcomes.

What started as a summer project became a profound learning experience. Samarth and his friends didn't just gain extra gaming hours; they learned invaluable life lessons. They discovered that success isn't always about grand gestures or breakthroughs but often about small, consistent actions that compound over time. They learned the power of consistency, the beauty of perseverance, and the joy of transforming something they love into something that helps them grow. This, my friends, is what it means to succeed. It's not about the destination but the journey and the growth that happens along the way. Samarth completed his internship, gaining real-life experience in web design, content creation, authoring, AI implementation, and leadership development. The future of innovation and leadership is in good hands.

Building Consistency

Consistency, the secret ingredient in the recipe for success, resonates through the actions of the world's most influential leaders.

Consistency is underpinned by discipline—the unwavering commitment to show up, work, and stay the course even when inspiration seems to be on hiatus. The outcome is an unshakeable foundation of perseverance and determination that continues to support your growth, regardless of the circumstances.

Consistency also requires resilience, the ability to turn setbacks into comebacks. On this journey, you'll encounter challenges and obstacles. However, the ability to bounce back, learn from these experiences, and persist leads to steady and consistent growth.

Authenticity is another crucial pillar. Staying true to your values and purpose fosters trust and connection with those around you and maintains your integrity. The result becomes a reputation built on transparency and genuine intent.

Patience plays a significant role, too. The most outstanding achievements don't happen overnight but result from consistent effort. Tolerance ensures you remain steadfast, even when immediate results seem elusive, leading to long-term success.

Embracing discomfort is a vital part of consistency. Venturing outside your comfort zone is where growth happens. The outcome is a more potent, more adaptable you, ready to take on whatever comes your way.

Accountability keeps you on track, ensuring your actions align consistently with your goals. Alongside a commitment to continuous learning and value addition, it fuels your progress toward your objectives, resulting in a sense of accomplishment and momentum.

Precise goal setting provides direction, offering measurable milestones to assess your progress. Coupled with adaptability, it allows you to navigate life's ever-changing landscape while staying true to your core values and objectives.

Remember, the essence of consistency isn't about perfection. It's about steadily progressing toward your goals, learning from setbacks, and remaining committed to your journey. This yields a proven track record of resilience, growth, and unwavering dedication to achieving your dreams. Let's embark on this journey together, exploring consistency's power and potential.

ACTIVITY: Journey to Consistency and Unleashing Your Greatness

Welcome to an enlightening journey on cultivating consistency in your life. The goal is to create a customized program that suits your needs and lifestyle, helping you build discipline, resilience, authenticity, patience, and more.

STEP 1: Assess Your Current Level

For each element of consistency, we provided three levels of activities: easy, medium, and advanced. Start by assessing where you currently stand for each element. Be honest with yourself; this is not about where you want to be but where you are right now.

STEP 2: Choose Your Activities

Based on your current level, pick one activity from each element. If you're at the easy level for discipline but advanced for resilience, that's perfectly okay. The aim is to create a balanced program that challenges you without being overwhelmed.

Let's delve into unique activities for each element to build consistency with concrete outcomes and timelines. For each component, we provided three levels of activities: easy, medium, and advanced. Pick from these or utilize them as inspiration for something else that comes to your mind. Emphasis is on building your consistency muscle.

1. Discipline:
    a. Easy: Start by making your bed every morning.
    b. Medium: Set aside 15 minutes daily for reading or personal development.

   c. Advanced: Commit to a 30-day challenge of a substantial task related to your work or personal growth.
2. Resilience:
   a. Easy: Practice mindful breathing for five minutes daily.
   b. Medium: Increase mindfulness meditation to 15 minutes daily.
   c. Advanced: To develop emotional resilience, begin journaling about your experiences and emotions.
3. Authenticity:
   a. Easy: Reflect on your values and write them down.
   b. Medium: Create a personal mission statement.
   c. Advanced: Review and reflect on your mission statement each morning.
4. Patience:
   a. Easy: Adopt a new hobby that requires learning a new skill.
   b. Medium: Choose a hobby that requires time to see results, such as gardening.
   c. Advanced: Engage in long-term projects like writing a book or building furniture.
5. Embrace Discomfort:
   a. Easy: Try a new food or recipe.
   b. Medium: Speak up in a meeting or public forum.
   c. Advanced: Regularly step out of your comfort zone, like public speaking or asking for feedback.
6. Accountability:
   a. Easy: Share your goals with a friend or family member.
   b. Medium: Regularly update them on your progress.
   c. Advanced: Recruit an accountability partner to keep you on track.

7. Continuous Learning:
    a. Easy: Listen to a podcast or watch a TED talk on a new topic each week.
    b. Medium: Dedicate 15 minutes daily to learning something new.
    c. Advanced: Enroll in an online course or start a new certification.
8. Value Addition:
    a. Easy: Compliment or thank someone each day.
    b. Medium: Find ways to add value to your work or relationships daily.
    c. Advanced: Start a project that adds significant value to your community or organization.
9. Goal Setting:
    a. Easy: Set a small, achievable goal for the week.
    b. Medium: Set specific, measurable, achievable, relevant, and time-bound (SMART) goals for the month.
    c. Advanced: Set and review SMART goals for the next three months.
10. Adaptability:
    a. Easy: Change one element in your daily routine.
    b. Medium: Try a new workout or morning routine each month.
    c. Advanced: Change a significant aspect of your life or work, like starting a new job or moving to a new city.

STEP 3: Document Your Program

Create a document to track your progress. It could be a simple spreadsheet or a dedicated journal. Write down the chosen activity for each element, and set specific, measurable goals. For example, if you're at the medium level for discipline, your plan might be to read a particular book for 15 minutes daily.

STEP 4: Set Your Timeline

Decide on a timeline for your program. A month is a good starting point, giving you enough time to see progress and allowing for adjustments if needed.

STEP 5: Track Your Progress

Update your document regularly with your progress. Did you make your bed every morning? Did you meditate for 15 minutes daily? Tracking your progress keeps you accountable and provides a visual representation of your journey toward consistency.

STEP 6: Reflect and Adjust

At the end of your timeline, reflect on your progress. Celebrate your successes, no matter how small they might seem. If some elements didn't go as planned, that's okay, too. The goal is not perfection but progress. Adjust your activities or goals as needed for the next cycle of your program.

Remember, consistency isn't about perfection. It's about making steady progress toward your goals, learning from your mistakes, and never giving up on your journey. In the grand scheme of life, consistency is not to be mistaken for stubbornness. It's a virtue—a measure of your steadfastness in your beliefs and values—not a mark of inflexibility. It's admirable when one can shift gears, adjust their viewpoint, and embrace a new perspective when needed. Just as a software engineer might switch from Python to JavaScript based on the project requirements, it helps you to pivot based on your firmly established values and nonnegotiable principles.

Raising your voice and demanding fairness might feel satisfying when an injustice comes your way. It's akin to a programmer finding a bug in the code—the immediate instinct is to fix it. However, I have learned over time that deviating from my consistent behavior, reacting impulsively, often leads to more harm than good.

Imagine working on a complex coding problem. You're frustrated, tired, and suddenly tempted to take a shortcut. But shortcuts often lead to more issues down the line, don't they? The same principle applies to life. Holding back and maintaining your composure amid negativity is challenging but crucial for long-term success.

You see, I have faith in you. I recognize your potential for greatness. This isn't empty flattery or blind optimism. My genuine concern for your well-being prompts me to make this personal appeal. Safeguard your peace of mind and wellness. Stand firm in your consistent behavior, rooted in your core values. Just as a seasoned tech professional would never compromise on quality, don't compromise on your principles for fleeting satisfaction. Instead, strive for the long-term fulfillment that comes from staying true to your ethos.

"Without commitment, you'll never start. But more importantly, without consistency, you'll never finish."
—Denzel Washington

You'll encounter difficult situations and people, but with consistency, you'll learn to maintain your integrity and protect your interests while preserving those of others. By consistently building physical and mental strength, you're helping yourself, your family, your community, and the world. Consistency allows others to feel safe with you, even if they disagree. Remember, the most enormous trees take the longest time to grow. Let's embark on this journey together, exploring the transformative power of consistency.

SECTION 2: BUILDING HABITS

Navigating the Labyrinth of Habit-Building in a Tech Professional's Career

In the dynamic world of technology, professionals face unique challenges and opportunities when it comes to habit-building for career development. This journey is often marked by frustrations and fears, fueled by desires, and guided by dreams. The pathway is complex, from maintaining productive work habits to fostering inclusivity, from managing professionalism amid challenges to continuous learning. The following table provides a deeper understanding of these experiences, highlighting 10 key areas. It captures the frustrations, desires, fears, and dreams that tech professionals encounter as they strive to build effective habits for their career growth.

Become a Professional with Robust Work Habits

The aspiration to become an exemplary professional with robust work habits is a dream that many tech professionals harbor. However, the journey is often punctuated by several challenges, including difficulty in maintaining a rhythm of productive work habits, a yearning to cultivate habits that amplify productivity, and anxiety about falling into unproductive routines. Despite these obstacles, success is achievable with a strategic and itemized plan. This guide presents a systematic approach to help tech professionals build and sustain habits that will lead them to their desired professional pinnacle:

1. Establish Clear Goals (Week 1-2): Start by defining what being an exemplary professional means to you. What work habits do you need to develop? Be specific and ensure your goals are SMART.

	Frustrations[35,39]	Desires[36,37,38]
1	Difficulty in maintaining a rhythm of productive work habits	Yearning to cultivate habits that amplify productivity
2	Struggle to accept and implement constructive feedback	Wish to improve at receiving and acting on criticism
3	Encountering bias within the tech industry	Want to help nurture an inclusive and diverse workplace
4	The challenge of correlating hard work with tangible results	Longing for recognition and reward for their efforts
5	Managing professionalism amid challenging situations	Hunger to maintain positivity even during difficult times
6	Tackling career derailers like conflict avoidance	Ambition to handle conflicts assertively and effectively
7	Overcoming the hurdle of continuous professional development	Passion for ongoing learning and skill enhancement
8	Frustration with the job-hunting process	Craving for a more streamlined job-hunting experience
9	Difficulty in fostering team player skills	Desire to be an integral part of a team
10	Struggle to maintain curiosity and the drive to keep learning	Yearning to keep their curiosity alive and continue learning

Fears[40]	**Dreams**[37,38]
Anxiety about falling into unproductive routines	Aspiration to be an exemplary professional with robust work habits
Worry of being resistant to feedback, hindering growth	The vision of becoming a team player who embraces criticism for growth
Fear of being sidelined due to prevailing biases	Dream of helping eliminate biases and stereotypes within tech
Dread of becoming overworked and underappreciated	Hope for a balanced life where effort correlates with reward
The concern of letting frustrations cloud their professionalism	Aim of harnessing emotional intelligence to navigate workplace hurdles
Apprehension that conflicts could lead to career derailment	The goal is to evolve into a leader skilled in conflict resolution
Fear of professional growth stagnating	Dream of being a lifelong learner, continually advancing in their career
Worry about settling for unsatisfying roles due to a challenging job market	The vision is to land fulfilling roles that align with their skills and passions
Fear of isolation due to lack of teamwork skills	Dream of being acknowledged as a valuable team contributor
The anxiety of losing the drive to learn and innovate	The hope of driving groundbreaking innovations through sustained curiosity

2. Create a Daily Routine (Week 3-4): Based on your defined goals, create a daily routine that incorporates activities that move you closer to these goals. This routine will help you maintain a productive rhythm and prevent you from falling into unproductive routines.
3. Start Small (Week 5-6): Begin with small, manageable tasks that contribute to your larger goal. This could be as simple as dedicating the first hour of your workday to the most challenging tasks or setting aside regular time for learning and skills development.
4. Consistency Over Intensity (Week 7 onward): It's better to consistently perform small actions toward your goal than to intensely pursue it sporadically. Consistency breeds habits, and over time, these habits will become second nature.
5. Regular Self-Evaluation (Every month): Regularly evaluate your progress. Are your actions leading you toward your goals? If not, adjust accordingly. The path to success is not always linear.
6. Overcoming Anxiety (Ongoing): Remember that everyone experiences unproductive days. Instead of succumbing to anxiety, learn from these experiences and use them as stepping stones toward your goal.
7. Seek Support (Ongoing): Don't hesitate to seek support when needed. Whether it's from a mentor or a peer, external perspectives can provide valuable insights and encouragement.

By following this strategic plan, you can expect to see gradual improvement in your work habits over time. You will become more disciplined and productive. This journey will not only bring you closer to becoming an exemplary professional but also instill a sense of confidence and fulfillment in your work.

Become a Team Player Who Embraces Criticism for Growth

For many tech professionals, the aspiration to become a team player who embraces criticism for growth is a highly sought-after dream. However, the path to achieving this goal can be fraught with challenges, such as struggling to accept and implement constructive feedback, the desire to improve at receiving and acting on criticism, and the fear of being resistant to feedback that could hinder personal and professional growth. To overcome these obstacles, a strategic and itemized plan is essential. This guide provides a roadmap to help tech professionals develop effective habits that will lead to success in becoming a team player who thrives on constructive criticism:

1. Embrace Feedback (Week 1-2): Start by acknowledging the value of feedback. Understand that feedback, especially constructive criticism, is an opportunity for growth and not a personal attack.
2. Seek Feedback Actively (Week 3-4): Cultivate the habit of asking for feedback from your peers, leaders, and juniors. This will help you become accustomed to receiving criticism and also shows your commitment to personal development.
3. Practice Reflective Listening (Week 5-6): When receiving feedback, listen attentively and reflect on what's said before responding. This practice will help you understand the crux of the feedback and how you can implement it.
4. Implement Feedback (Week 7 onward): Develop an action plan to implement the feedback received. This could involve learning new skills, adjusting your work style, or changing your approach to teamwork.
5. Regular Self-Evaluation (Every month): Evaluate your progress regularly. Are you improving in accepting and implementing feedback? If not, identify the barriers and find ways to overcome them.

6. Overcoming Fear (Ongoing): Everyone struggles with criticism. Instead of fearing it, view it as a tool for improvement. Use your fear as motivation to grow and improve.
7. Seek Support (Ongoing): Don't hesitate to seek support when you find it difficult to accept or implement feedback. A mentor or coach can provide valuable guidance and encouragement.

By following this strategic plan, tech professionals can expect to become more receptive to criticism, improve their teamwork skills, and experience personal and professional growth. Overcoming the fear of criticism and learning to use it as a tool for improvement will not only contribute to becoming an effective team player but also lead to increased self-confidence and job satisfaction.

Eliminate Biases and Stereotypes Within Tech

The dream of eliminating biases and stereotypes within the tech industry is a noble aspiration that many tech professionals share. However, this journey can be marked by frustrations, such as bias within the industry, a strong desire to nurture an inclusive and diverse workplace, and fear of being sidelined due to prevailing biases. With a strategic and itemized plan, these obstacles can be overcome. This guide aims to help tech professionals develop habits and strategies that will lead them toward achieving this dream:

1. Educate Yourself (Week 1-2): Start by educating yourself about biases and stereotypes. Understand their origins, impacts, and ways to counteract them.
2. Promote Awareness (Week 3-4): Share your knowledge with your team and colleagues. Encourage open conversations about biases and stereotypes in the workplace.
3. Implement Inclusive Practices (Week 5-6): Begin incorporating inclusive practices into your work routine. This could

include advocating for diverse hiring practices or ensuring all voices are heard in meetings.

4. Encourage Feedback and Open Communication (Week 7-8): Foster an environment where colleagues feel safe to express concerns about biases and stereotypes. Regularly ask for feedback and be open to criticism.
5. Champion Diversity (Month 2 onward): Actively support diverse colleagues and stand up against bias and discrimination. Be the change you wish to see in your workplace.
6. Continual Learning (Ongoing): Stay updated with the latest research and trends related to diversity and inclusion. Attend workshops, webinars, and conferences to expand your knowledge.
7. Resilience in the Face of Resistance (Ongoing): Not everyone may agree with your efforts. Be prepared for resistance and stay resilient. Change takes time.

By following this strategic plan, you can expect to become a strong advocate for diversity and inclusion in your workplace. You will encourage others to recognize and challenge their own biases, fostering a more inclusive environment. Over time, these actions can contribute to a significant shift in the workplace culture, helping to eliminate biases and stereotypes within the tech industry.

Balance Life Where Effort Correlates with Reward

Many tech professionals dream of a balanced life, where effort correlates directly with reward. However, this dream can often be overshadowed by frustrations, such as the challenge of correlating hard work with tangible results, a longing for recognition and reward for their efforts, and the dread of becoming overworked and underappreciated. With a well-crafted, strategic plan, these challenges can be overcome. This guide provides a roadmap to help tech professionals build habits that will lead them toward achieving their dream:

1. Define Your Definition of Success (Week 1-2): Start by clearly defining what success means to you. Is it a promotion, a salary increase, or perhaps achieving certain milestones? Having a clear definition will help you align your efforts with your goals.
2. Set SMART Goals (Week 3-4): Based on your definition of success, set SMART goals. These goals will guide your efforts and provide a benchmark for measuring your progress.
3. Establish a Productive Routine (Week 5-6): Develop a daily routine that supports your goals. This could include prioritizing tasks, managing time effectively, and taking regular breaks to avoid burnout.
4. Communicate Your Efforts (Week 7-8): Don't be shy about sharing your accomplishments with your superiors. Regularly update them about your progress and the efforts you contribute.
5. Seek Feedback and Recognition (Month 2 onward): Actively seek feedback on your work. Constructive feedback can help you improve, while positive feedback can serve as recognition of your efforts.
6. Practice Self-Care (Ongoing): Remember that a balanced life includes taking care of your physical and mental health. Incorporate self-care habits into your routine, such as regular exercise, a healthy diet, and adequate sleep.
7. Resilience in the Face of Challenges (Ongoing): There will be times when your efforts may not seem to yield the desired results. Stay resilient, learn from these experiences, and keep striving toward your goals.

By following this strategic plan, tech professionals can expect to see a stronger correlation between their efforts and rewards. They will feel more recognized for their work and experience a better work-life balance. This will lead to increased job satisfaction, improved well-being, and a sense of fulfillment in their professional life.

Harness Emotional Intelligence to Navigate Workplace Hurdles

The dream of harnessing emotional intelligence to navigate workplace hurdles is a common goal among tech professionals. However, this journey can sometimes be marked by frustrations, such as managing professionalism amid challenging situations, a hunger to maintain positivity even during difficult times, and the concern of letting frustrations cloud their professionalism. This guide aims to help tech professionals develop habits and strategies that will lead them toward achieving this dream:

1. Understand Emotional Intelligence (Week 1-2): Start by educating yourself about emotional intelligence. Understand its five components—self-awareness, self-regulation, motivation, empathy, and social skills.
2. Develop Self-Awareness (Week 3-4): Begin cultivating self-awareness. Practice mindfulness, be aware of your emotions, and understand how they affect your actions.
3. Practice Self-Regulation (Week 5-6): Learn to control your emotions, especially in stressful situations. Techniques such as deep breathing or taking a moment to compose yourself can be helpful.
4. Cultivate Empathy (Week 7-8): Develop empathy by actively listening to others and trying to understand their perspectives. This can help you navigate interpersonal relationships in the workplace more effectively.
5. Improve Social Skills (Month 2 onward): Work on your social skills. Learn how to communicate effectively, manage conflicts, and work well in teams.

6. Stay Motivated (Ongoing): Keep your motivation levels high, even in challenging situations. Remember your goals and why you strive to achieve them.
7. Seek Feedback and Continual Learning (Ongoing): Regularly seek feedback on your emotional intelligence skills. Use it as a way to continually learn and improve.

By following this strategic plan, tech professionals can expect to develop strong emotional intelligence skills. They will be better equipped to manage challenging situations, maintain positivity, and uphold professionalism. These skills will not only help them navigate workplace hurdles more effectively but also lead to improved relationships with colleagues, job satisfaction, and overall career success.

Evolve Into a Leader Skilled in Conflict Resolution

The aspiration to evolve into a leader skilled in conflict resolution is a commendable goal for any tech professional. This journey, however, can be fraught with frustrations like tackling career derailers such as conflict avoidance. It comes with the ambition to handle conflicts assertively and effectively but also the apprehension that conflicts could lead to career derailment. This guide aims to provide a strategic and itemized plan to help tech professionals cultivate habits that will lead them to success in accomplishing this dream:

1. Understand Conflict Resolution (Week 1-2): Begin by educating yourself about conflict-resolution strategies. Understand the root causes of conflicts and the techniques to resolve them effectively.
2. Develop Empathy (Week 3-4): Cultivate empathy—a key component in conflict resolution. Strive to understand the perspectives of others and validate their feelings.

3. Practice Active Listening (Week 5-6): Enhance your active listening skills. This involves not just hearing what is said but also understanding the underlying emotions and concerns.
4. Improve Communication Skills (Week 7-8): Work on your communication skills. Learn to express your thoughts and concerns clearly and assertively, without resorting to aggression or passivity.
5. Apply Conflict-Resolution Techniques (Month 2 onward): Start applying conflict-resolution techniques in real-life situations. Use strategies such as mediation, negotiation, and problem-solving.
6. Seek Feedback and Continual Learning (Ongoing): Regularly seek feedback on your conflict-resolution skills. Use it as a way to continually learn and improve.
7. Resilience in the Face of Challenges (Ongoing): Be prepared for setbacks. Not all conflicts may be resolved easily or quickly. Stay resilient and committed to your goal.

By following this strategic plan, tech professionals can expect to develop strong conflict-resolution skills. They will be better equipped to handle conflicts assertively and effectively, preventing potential career derailment. These skills will not only help them navigate workplace conflicts more effectively but also position them as effective leaders in their respective fields.

Become a Lifelong Learner, Continually Advancing in Your Career

The dream of becoming a lifelong learner and continually advancing in one's career is a fundamental ambition for tech professionals. However, this journey can often be marred by frustrations, such as the hurdle of continuous professional development, a passion for ongoing learning and skill enhancement, and the fear of professional growth stagnating. This guide aims to help tech professionals develop habits that will lead them toward achieving this dream:

1. Identify Learning Opportunities (Week 1-2): Begin by identifying areas for growth and potential learning opportunities. This could include new technologies, programming languages, or project management methodologies.
2. Set Learning Goals (Week 3-4): Define specific learning goals based on the opportunities you have identified. Make them achievable, clear, and set milestones.
3. Create a Learning Schedule (Week 5-6): Develop a regular learning schedule. This could involve dedicating a certain number of hours each week to learning. Ensure the schedule is realistic and manageable.
4. Engage in Active Learning (Week 7 onward): Start actively pursuing your learning goals. This could involve reading books, taking online courses, attending webinars, or practicing new skills.
5. Seek Feedback and Apply Learning (Month 2 onward): Regularly seek feedback on your progress and apply what you learned in real-life situations. This will reinforce your learning and demonstrate your growth.
6. Keep Up with Industry Trends (Ongoing): Stay abreast of industry trends and developments. This will help you identify new learning opportunities and stay relevant in your field.
7. Maintain a Growth Mindset (Ongoing): Cultivate a growth mindset, viewing challenges as opportunities to learn and grow rather than obstacles.

By following this strategic plan, tech professionals can expect to become lifelong learners, continually advancing in their careers. They will overcome the hurdles of continuous professional development and alleviate the fear of professional growth stagnating. This continual learning and advancement will lead to increased job satisfaction, career opportunities, and a sense of accomplishment.

Land Fulfilling Roles that Align with Your Skills and Passions

The vision to land fulfilling roles that align with one's skills and passions is a common goal among tech professionals. However, this dream can often be overshadowed by frustrations, such as the tedious job-hunting process, a craving for a more streamlined experience, and the worry about settling for unsatisfying roles due to a challenging job market. This guide aims to help tech professionals develop habits that will lead them toward achieving this dream:

1. Identify Your Skills and Passions (Week 1-2): Start by identifying your skills and passions. Understand what you're good at and what you enjoy doing. This will guide you in your job hunt.
2. Research Potential Roles (Week 3-4): Look for roles that align with your skills and passions. Use job search engines, network with industry professionals, and attend career fairs or meetups.
3. Tailor Your Applications (Week 5-6): Customize your résumé and cover letter for each application. Highlight your relevant skills and experiences and convey your passion for the role.
4. Prepare for Interviews (Week 7-8): Practice common interview questions and prepare examples to demonstrate your skills and experiences. Also, research the company and role thoroughly.
5. Maintain a Positive Mindset (Ongoing): Job hunting can be challenging. Maintain a positive mindset and remember that rejections are part of the process.

6. Continual Skill Development (Ongoing): Keep improving your skills and knowledge. Stay updated with industry trends and advancements.
7. Network Regularly (Ongoing): Networking is key in the job hunt. Regularly connect with industry professionals, attend relevant events, and leverage social media platforms like LinkedIn.

By following this strategic plan, tech professionals can expect to navigate the job-hunting process more effectively and land roles that align with their skills and passions. This will reduce frustrations with the job-hunting process, provide a more streamlined experience, and alleviate worries about settling for unsatisfying roles. Ultimately, this will lead to increased job satisfaction and career fulfillment.

Receive Acknowledgment as a Valuable Team Contributor

The dream of being acknowledged as a valuable team contributor is a common ambition among tech professionals. However, this aspiration can often be overshadowed by difficulties in fostering team player skills, the desire to be an integral part of a team, and the fear of isolation due to a lack of teamwork skills. This guide aims to help tech professionals create habits that will lead them toward achieving this dream:

1. Identify Key Team Player Skills (Week 1-2): Begin by identifying the key skills that make a good team player, such as communication, collaboration, reliability, and adaptability.
2. Set Personal Goals (Week 3-4): Based on the identified skills, set personal development goals. Use the collaborative, limited, emotional, appreciable, and refinable (CLEAR) goal framework.

3. Improve Communication Skills (Week 5-6): Start working on your communication skills, which are vital for effective teamwork. Learn to listen actively, express yourself clearly, and provide constructive feedback.
4. Develop Collaboration Skills (Week 7-8): Focus on strengthening your collaboration skills. Get involved in team projects, contribute ideas, and learn to compromise when necessary.
5. Demonstrate Reliability (Month 2-3): Show your team that you're reliable. Meet deadlines, keep your commitments, and be consistent in your performance.
6. Cultivate Adaptability (Month 4 onward): In the dynamic tech world, adaptability is key. Be open to new ideas, willing to learn, and ready to adjust to changing circumstances.
7. Seek Feedback and Improve (Ongoing): Regularly seek feedback from your team members and superiors. Use it to identify areas for improvement and track your progress.

By following this strategic plan, tech professionals can expect to enhance their team player skills and become valuable contributors to their teams. They will overcome difficulties in fostering teamwork skills, satisfy their desire to be integral team members, and alleviate fears of isolation. This will lead to increased job satisfaction, improved team performance, and recognition as a valuable team contributor.

Drive Groundbreaking Innovations Through Sustained Curiosity

The aspiration to drive groundbreaking innovations through sustained curiosity is a noble and exciting ambition for tech professionals. However, the path to achieving this can often be hindered by frustrations, such as the struggle to maintain curiosity and the drive to keep learning, the yearning to keep their curiosity alive and continue learning, and the anxiety of losing the drive to learn and innovate. This guide aims to help tech professionals establish habits that will lead them toward achieving this dream:

1. Cultivate a Curious Mindset (Week 1-2): Start by cultivating a curious mindset. Keep asking questions, exploring new ideas, and challenging assumptions. This is the fuel for innovation.
2. Create a Learning Plan (Week 3-4): With your curiosity sparked, create a learning plan. Set actionable learning targets and paths.
3. Engage in Lifelong Learning (Week 5-6): Commit to lifelong learning. Dedicate a specific amount of time each day to learning something new, whether it's reading, taking online courses, or experimenting with new technologies.
4. Develop a Habit of Reflection (Week 7-8): Regularly reflect on what you learned and how it can be applied to drive innovation. This helps to consolidate learning and sparks new ideas.
5. Seek Diverse Perspectives (Month 2-3): Expand your horizons by seeking diverse perspectives. Connect with different people, attend varied events, and explore other industries. This can fuel creativity and innovation.
6. Embrace Failure as a Learning Opportunity (Month 4 onward): Don't be discouraged by failure. Instead, see it as an opportunity to learn and improve. This helps to maintain the drive to innovate.
7. Regularly Review and Adjust Your Learning Plan (Ongoing): As you grow and evolve, so should your learning plan. Regularly review and adjust it to ensure it remains relevant and challenging.

By following this strategic plan, tech professionals can expect to sustain their curiosity, fuel their drive to learn, and foster a culture of innovation. They will overcome the struggle to maintain curiosity, satisfy their yearning to continue learning, and alleviate the anxiety of losing the drive to innovate. This will lead to personal growth, groundbreaking innovations, and a fulfilling and exciting career in tech.

There's an invigorating journey ahead of you, a journey where your dreams are within reach and your ambitions are not just figments of your imagination but achievable milestones. Imagine a world where you're not just surviving but thriving in your tech career, where you're not just another face in the crowd but an invaluable team player, a lifelong learner, and a driving force for groundbreaking innovation.

Imagine the thrill of identifying key skills, setting personal goals, improving communication, and developing collaboration skills. Picture yourself demonstrating reliability, cultivating adaptability, and continuously seeking feedback to improve. This is the reality of becoming a valuable team contributor.

Next, envision yourself as a lifelong learner, continually advancing in your career. You'll identify learning opportunities, set clear goals, schedule regular learning sessions, and actively engage in acquiring new knowledge. You'll apply your learning in real-world situations, keep up the pace with industry trends, and cultivate a growth mindset.

Finally, imagine driving groundbreaking innovations through sustained curiosity. You'll cultivate a curious mindset, create a learning plan, engage in lifelong learning, and develop a habit of reflection. You'll seek diverse perspectives, embrace failure as a learning opportunity, and regularly review and adjust your learning plan.

Each step of these strategic plans is a stepping stone toward a more fulfilling, rewarding, and successful career in tech. It's not about the destination but the journey. It's about the habits you build, the skills you acquire, and the person you become.

So, why wait? The time for action is now. Embrace these plans, give them an honest try, and watch as they transform not only your career but also your life. Remember, success is something you attract by the person you become. So, go ahead, seize the day, and embark on this exciting journey of personal and professional growth.

SECTION 3: PRACTICE BY TEACHING OTHERS

When we really dig into our knowledge—when we dissect it and flip it inside out to explain it to someone else—we gain a deeper understanding of what we know. It's like trying to teach someone how to swim. You've been doing it all your life, but when you have to break down the mechanics of it to someone else, you truly grasp the nuances of each stroke.

Let's think about vulnerability for a moment. When we share our stories, experiences, and wisdom, we are vulnerable. We step into the arena without armor. In doing so, we not only empower others with our insight, but we also reinforce it within ourselves. It's like sharing your favorite song with a friend—you relate to it on one level, but explaining why it touches you adds another layer of connection to it.

We grow when we share. It's a two-way street. Imagine a tree, its roots spreading out, reaching others and intertwining. Every time we share, we extend our roots, creating a network of growth and learning. We're not just helping others grow; we're growing ourselves.

The beauty of sharing is that it expands us. Think of your heart as a suitcase. The more you pack into it, the more it can hold. When you share your wisdom, you're not emptying your suitcase; you're making it bigger. You're creating more room for experiences, ideas, and emotions.

However, not everyone can pack the same amount into their suitcase. Some people haven't stretched their capacity yet. They haven't learned how to unpack and repack, to let go of the old to make room for the new. And that's okay. Because every time we share, we give them the opportunity to learn how. We show them how to expand their suitcase.

Sharing is a path to growth. It allows us to hold and experience more. The richer our experiences, the deeper our understanding and the greater our happiness. It's like a beautiful dance, where comprehension and joy move in sync with each other, each step expanding our capacity for the next.

So, share your wisdom, stories, and experiences, not just for others but for you. Every time you share, you step onto the dance floor of growth. You expand your heart's suitcase. You become bigger.

Focus on the Path, Not the Trees: The Role of Humility and Gentleness in Teaching

Indeed, it's a real scenario for someone to come across as boastful or overbearing while teaching. This usually arises when the person doing the teaching lacks a touch of humility or doesn't consider the knowledge and viewpoint of the individual they instruct.

Effective teaching is like a two-lane highway—it should involve as much active listening and empathy as sharing knowledge. Teaching isn't about showcasing your supremacy or dominance but about assisting others in their journey of learning and growth. It's essential to cultivate an atmosphere of mutual respect, where questions are welcomed and everyone's viewpoints are treasured.

When you're in the role of a teacher, be mindful of your tone, language, and approach. For example, using straightforward language, providing examples from real life, and demonstrating patience can significantly impact how the information is perceived. Also, by asking for feedback and being open to learn from those you teach, you can avoid the trap of appearing arrogant or domineering.

Let's not forget the objective of teaching isn't just to distribute what you know but to ignite a passion for learning in others. Consider this analogy: hiring a ski instructor. You'd expect them not to teach you to dodge the trees but, instead, to focus on the path. When you tell someone to evade the trees, all they see are trees and they're more likely to collide with one. But when the instructor tells you to concentrate on the path, that's all you see. The probability of hitting a tree drops significantly. In the same vein, I could tell you not to boast and not to impose, but then, I'm telling you to dodge the trees instead of focusing on the path. To stay on the path, embrace humility and gentleness when sharing what you know. Let's delve more into this.

When we're learning something new—like skiing—we need to fix our attention on where we aspire to go, not what we seek to avoid. The same rule applies when disseminating our knowledge with others. When we embrace teaching from a place of humility and gentleness, we naturally veer away from boasting or imposing our thoughts. Humility means accepting that while we may be experts in a particular subject, we don't know it all—and that's perfectly fine. There's always room for growth and learning, even for teachers.

Humility enables us to listen and be receptive to the perspectives of those we instruct. It promotes dialogue and nurtures an environment of mutual respect and understanding. When we're humble, we're more engrossed in the learning process, rather than validating how much we know.

Gentleness in our approach is another crucial element in effective teaching. It's not solely about what we articulate but how we convey it. Gentle teaching involves patience, understanding, and empathy. We need to respect the pace at which the other person learns and adapt our teaching style to suit them.

Being gentle also signifies being mindful of our language and tone. Our aim should be to encourage and inspire, not criticize or belittle. Our goal is to guide the learner on their path of discovery, not shove them toward it.

Keeping focused on the path—being humble and gentle—when sharing our knowledge can significantly enrich the learning experience for both the teacher and the learner. It fosters a positive learning environment where ideas can be freely exchanged and probed, leading to deeper comprehension and growth for everyone involved.

The Spectrum of Learners: Insights from Abisoye Ajayi-Akinfolarin's Educational Transformation Journey

Abisoye Ajayi-Akinfolarin, a social impact entrepreneur and founder of the Pearls Africa Foundation, has a distinctive approach to educational transformation. This is particularly evident in her work with disadvantaged girls in Lagos, Nigeria, where she empowers them through technology education.

First, Ajayi-Akinfolarin recognizes that not every girl will have access to or embrace technology education due to socioeconomic constraints. However, she doesn't let this deter her efforts. Instead, she focuses on those who are open to learning and can benefit from the value that technology education offers, much like Goodall's approach with her conservation efforts,[41]

Ajayi-Akinfolarin also plays a crucial role in enlightening these girls about the gaps in their knowledge and demonstrating how technology education can fill those gaps. This requires a deep understanding of their needs and desires, which she actively works toward understanding and addressing.

There are those who acknowledge their need for technology education but don't see it as an immediate necessity due to other pressing issues in their lives. Here, Ajayi-Akinfolarin differentiates her education program by making it not just about imparting technological skills but also providing the girls with tools to become financially independent. This adds unique value to her program and makes it indispensable.

Lastly, there are girls who are ready and eager to learn from Ajayi-Akinfolarin right now. These are her most promising students. However, she doesn't take their interest for granted. She ensures her teaching approach is as engaging and remarkable as the knowledge she imparts, making technology education accessible and engaging.

In essence, Ajayi-Akinfolarin's approach to educational transformation is about understanding needs, educating, differentiating her program, and providing a compelling learning experience. By applying these principles, she has successfully empowered many young girls in Nigeria, transforming their lives through education.

Harnessing Digital Platforms: A New Frontier for Tech Professionals to Educate and Inspire

In today's digital age, there are numerous platforms and resources for tech professionals to leverage in order to share their knowledge. Here are some strategies to consider:

1. Leverage YouTube: YouTube is an excellent platform for delivering educational content. Creating tutorial videos or

webinars on specific tech topics can be an effective way to share your expertise. Ensure your videos are clear, concise, and engaging, with a good balance of theory and practical demonstration.

2. Utilize Social Media: Platforms like LinkedIn, Twitter, and Instagram aren't just for socializing. They're also powerful tools for sharing professional insights. Share short tips, industry news, or thought-provoking questions related to your field. Use relevant hashtags to reach a wider audience.
3. Start a Blog or Write on Medium: Writing blog posts or articles on Medium can help you delve deeper into topics and provide valuable information to readers. Plus, it helps establish you as an authority in your field. Make sure to promote your posts on your social media channels.
4. Host Webinars or Online Classes: Platforms like Zoom or Google Meet can be used to host webinars or online classes. These can be one-off sessions on specific topics or a series of classes covering a broader subject.
5. Publish an eBook or Create a Course: If you have extensive knowledge on a topic, consider compiling it into an eBook or an online course on platforms like Udemy or Coursera. This not only helps others learn but can also generate passive income.
6. Use Book Summaries or Podcasts: If you're more comfortable speaking than writing, consider starting a podcast. You can discuss various tech topics, interview other professionals, or even summarize key points from relevant books.
7. Engage in Online Communities: Participate in online forums and communities such as Reddit, Quora, or Stack Overflow. Answering questions and providing advice in these forums can help you share your knowledge with those seeking to learn.

The goal is to make learning convenient for your audience. So, focus on creating high-quality content that's accessible, easy to understand, and valuable. Don't forget to engage with your audience—respond to comments, answer questions, and solicit feedback. This will help you build a community around your content and increase its impact.

Regardless of your position in your day job, you can start a side project and spend a few hours weekly on any of these strategies. Sharing knowledge and expertise is not only rewarding, but it also contributes to the wider tech community. Here are a few points to keep in mind:

- Time Management: Balancing a full-time job and a side project can be challenging. It's important to plan your time effectively and set realistic goals for your side project.
- Conflict of Interest: Make sure your side project doesn't conflict with your primary job responsibilities or your company's policies. It's always a good idea to check with your HR department if you're unsure.
- Passion: Choose a platform or a strategy that aligns with your interests. If you're passionate about what you're doing, it won't feel like an additional chore.
- Consistency: Whether it's writing a blog post, making a YouTube video, or answering questions on forums, try to be consistent. Regular engagement helps build a loyal audience and increases your impact.
- Learning Curve: Each platform has a learning curve. For example, if you choose YouTube, you'll need to learn video editing. Be prepared to invest some time in learning these skills.
- Patience: Building an audience or a community takes time. Don't become discouraged if you don't see immediate results. Keep delivering valuable content, and your efforts will pay off over time.

The goal is to share your knowledge and contribute to the community. So, even a few hours a week can make a big difference.

ACTIVITY: Unleash Your Tech Expertise on YouTube: An Exciting and Accessible Journey

STEP 1: Platform Exploration (Day 1-3)

- Immerse yourself in the YouTube universe. Study and learn from successful tech channels.
- Expect: Understand what content clicks with viewers and identify ongoing trends.
- Don't be daunted by existing channels. Your unique take is your power.

STEP 2: Swift Content Production (Day 4-10)

- Step in front of the camera and start filming. Remember, completion trumps perfection. Refinement can happen as you progress.
- The initial attempts may not be flawless. With each video, your skills will improve.
- Don't let the quest for perfection slow you down. Speed is crucial.

STEP 3: Channel Configuration (Day 11)

- Establish your YouTube channel. Ensure your branding is uniform and professional.
- A well-branded channel will draw more viewers.
- Make sure your channel name is congruent with your content and personal brand.

STEP 4: Publish and Fine-Tune (Day 12)

- Release your first video. Make it more discoverable with a catchy title, detailed description, and relevant tags.
- Initial views might be modest, but consistency is the key to growth.
- Don't forget to choose the correct category for your video.

STEP 5: Consistent Content Delivery (Day 13 onward)

- Maintain a steady flow of video creation and uploads. The more content you produce, the quicker your growth.
- As you continue uploading, you will begin to see an increase in views and subscribers.
- Don't compromise quality for the sake of quantity. Each video should add value.

STEP 6: Community Engagement (Ongoing)

- Respond to all comments. Your viewers are your community—show them respect.
- Higher engagement results in better visibility on YouTube.
- Negative comments can be challenging, but they also provide opportunities for improvement.

The principle of "document, don't create" means you should concentrate on recording your expertise and insights, rather than trying to produce flawless content. So, gear up and start sharing your knowledge with the world.

ACTIVITY: Tech Professionals on Social Media: The Ultimate Guide to Building Your Online Presence

This tutorial will guide technology professionals to leverage social media platforms (LinkedIn, Twitter, and Instagram) for professional growth and engagement.

STEP 1: Understand Social Media Platforms (1 day)

- Explain the unique features of each platform. For instance, LinkedIn is excellent for professional networking and job hunting, Twitter for quick updates and thought leadership, and Instagram for visual content.
- Social media allows you to share insights, connect with peers, stay updated with industry trends, and even find new opportunities.
- Each platform has its own culture and etiquette. Observing how others interact can help you understand the norms.

STEP 2: Plan Your Social Media Presence (1-2 days)

- Identifying your goals. Are you aiming to network, share expertise, or learn from others? Clear goals will guide your strategy.
- Plan to share a mix of content types—articles, quick tips, industry news, personal updates about projects, etc. Using relevant hashtags can increase visibility.
- Consistency is key in social media. Consider your available time to decide on a sustainable posting schedule.

STEP 3: Set Up Your Profiles (1 day)

- Set up your account and profile with a professional email address.
- Optimize your profiles. Use a professional photo, write clear bios explaining what you do, and add relevant links (like your GitHub or portfolio).
- It might take some iterations to finalize your profile information. That's perfectly normal.

STEP 4: Create and Share Content (Ongoing)

- Keep your content informative, relevant, and engaging. Use a friendly tone that invites conversation. For visual platforms like Instagram, good aesthetics are crucial.
- Post during peak hours when your audience is most active. Use platform-specific features (like LinkedIn articles or Twitter threads) to share long-form content.
- Be respectful and avoid posting controversial or sensitive content.

STEP 5: Engage with Your Audience (Ongoing)

- Respond to comments on your posts and engage in conversations on others' posts. This can help you build relationships and increase your visibility.
- You can occasionally cross-promote your content across platforms. But make sure the content fits the platform where you're sharing it.
- Building an engaged following takes time and patience.

STEP 6: Monitor Your Progress (Weekly/Monthly)

- Use built-in analytics tools to understand which types of posts foster the most engagement.
- Based on your analysis, adjust your strategy. Maybe your audience prefers videos over text posts, or they're more active at a different time.
- Don't become discouraged by slow growth. It's better to have a small, engaged audience than a large, uninterested one.

Consistent use of social media can position you as a thought leader, expand your network, and open up new opportunities. Keep refining your strategy based on your performance and goals. As you grow, consider exploring advanced tactics like collaborations or sponsored posts.

Social media is a tool. The real value comes from the connections and conversations you have, not just the number of followers you accumulate.

Activity: Becoming a Tech Blogger is Easier Than You Think

This tutorial will guide technology professionals to start a blog or write on Medium for professional growth and thought leadership.

STEP 1: Understand Blogging Platforms (1 day)

- Study the unique features of blogging platforms like WordPress, Blogger, and Medium. For instance, Medium has a built-in audience, WordPress offers more customization, and Blogger is user-friendly for beginners.
- Blogging allows you to delve deeper into topics, share insights and research, and establish your authority in your field.

- Each platform has its own interface and writing tools. Spend some time getting familiar with these before you start writing.

STEP 2: Plan Your Blogging Strategy (1-2 days)

- Identify your goals. Are you aiming to educate others, share your expertise, or build a personal brand? Clear goals will guide your content strategy.
- Plan to write on topics that align with your expertise and interests. Also consider what your target audience finds valuable.
- Consistency is key in blogging. Consider your available time to decide on a sustainable posting schedule.

STEP 3: Set Up Your Blog (1-2 days)

- Based on your needs and comfort level, choose a blogging platform. Set up your account with a professional email address.
- Choose a clean, professional theme. Make sure it's easy for visitors to navigate and find posts.
- It might take some iterations to finalize your blog design. That's perfectly normal.

STEP 4: Write and Publish Your Posts (Ongoing)

- Keep your posts informative and engaging. Break up text with subheadings, bullet points, and images to make it easier to read.
- Once you're happy with your post, publish it. Make sure to proofread for any errors before hitting the publish button.

- Plagiarism is a serious offense in blogging. Always attribute sources if you're using someone else's work.

STEP 5: Promote Your Posts (Ongoing)

- Share your posts on your social media channels to reach a wider audience. Use relevant hashtags to increase visibility.
- Respond to comments on your posts and engage in conversations. This can help you build relationships and increase your readership.
- Building an engaged readership takes time and patience.

STEP 6: Monitor Your Progress (Weekly/Monthly)

- Use built-in analytics tools to understand which types of posts receive the most views and shares.
- Based on your analysis, adjust your strategy. Maybe your audience prefers tutorials over opinion pieces, or they're more engaged on certain topics.
- Don't become discouraged by slow growth. It's better to have a small, engaged audience than a large, uninterested one.

Consistent blogging can position you as a thought leader, expand your network, and open up new opportunities like speaking engagements or consulting gigs. Keep refining your strategy based on your performance and goals. As you grow, consider exploring advanced tactics like guest posting or monetization strategies.

Blogging is your own little corner of the Internet. The magic doesn't come from how many people stop by. It's all about the wisdom you sprinkle and the relationships you cultivate along this wonderful journey.

ACTIVITY: Amplify Your Influence: A Tech Professional's Roadmap to Podcasting for Thought Leadership

This tutorial will guide tech professionals on how to leverage podcasts for professional growth and thought leadership.

STEP 1: Understand the Platforms (1 day)

- Familiarize yourself with popular podcasting platforms such as Anchor, SoundCloud, Buzzsprout, and Podbean. Each platform has unique features and benefits that can be used to your advantage.
- Podcasting allows you to share insights, connect with peers, stay updated with industry trends, and even find new opportunities.
- Each platform has its own user interface and tools. Spend some time getting familiar with these before you start creating content.

STEP 2: Plan Your Content (1-2 days)

- Identify your goals. Are you aiming to educate others, share your expertise, or build a personal brand? Clear goals will guide your content strategy.
- Plan to create content that aligns with your expertise and interests. Also consider what your target audience finds valuable.
- Consistency is key in content creation. Consider your available time to decide on a sustainable posting schedule.

STEP 3: Set Up Your Account (1-2 days)

- Based on your needs and comfort level, choose a podcasting platform. Set up your account with a professional email address.
- Choose a clean, professional page and theme. Make sure it's easy for visitors to navigate and find your podcasts.
- It might take some iterations to finalize your channel/page design. That's perfectly normal.

STEP 4: Create and Publish Your Podcasts (Ongoing)

- Keep your podcasts informative and engaging. If you script your episodes, break up text with subheadings, bullet points, and images to make it easier to read.
- Once you're happy with your podcast, publish it. Make sure to listen for any errors before hitting the publish button.
- Plagiarism is a serious offense. Always attribute sources if you use someone else's work.

STEP 5: Promote Your Podcasts (Ongoing)

- Share your podcasts on your social media channels to reach a wider audience. Use relevant hashtags to increase visibility.
- Respond to comments on your podcasts and engage in conversations. This can help you build relationships and increase your audience.
- Building an engaged audience takes time and patience.

STEP 6: Monitor Your Progress (Weekly/Monthly)

- Use built-in analytics tools to understand which types of podcasts receive the most listens and shares.
- Based on your analysis, adjust your strategy. Maybe your audience prefers certain topics over others, or they're more engaged on certain platforms.
- Don't become discouraged by slow growth. It's better to have a small, engaged audience than a large, uninterested one.

Consistent podcasting can position you as a thought leader, expand your network, and open up new opportunities like speaking engagements or consulting gigs. Keep refining your strategy based on your performance and goals. As you grow, consider exploring advanced tactics like guest posting or monetization strategies.

Podcasting is like your own personal magnifying glass. However, the true treasure doesn't lie in the crowd you gather. It's in the secrets you unveil and the ties you create along the journey.

ACTIVITY: Tech Pro to Newsletter Ninja: Your Comprehensive Guide to Starting an Organization's Newsletter

This tutorial will guide tech professionals on how to start or contribute to their organization's monthly newsletter. The aim is to share technical, business, or best practices knowledge to increase influence, enhance leadership perception, and accelerate career growth.

STEP 1: Understand the Purpose of Newsletters (1 day)

- Understand the unique features of newsletters as a medium for communication. They're excellent for sharing in-depth insights, updates, and best practices within an organization.

- Newsletters allow you to share insights, stay updated with industry trends, and even establish yourself as a thought leader within your organization.
- Each organization has its own culture and etiquette around newsletters. Familiarize yourself with these norms.

STEP 2: Plan Your Newsletter Contribution (1-2 days)

- Identify your goals. Are you aiming to educate others, share your expertise, or build a personal brand? Clear goals will guide your content strategy.
- Plan to create content that aligns with your expertise and interests. Also consider what your colleagues find valuable.
- Consistency is key in content creation. Consider your available time to decide on a sustainable contribution schedule.

STEP 3: Set Up Your Newsletter Contribution Process (1-2 days)

- If your organization doesn't have a newsletter yet, identify a platform that suits your needs. Platforms like MailChimp or Substack are good options.
- Keep the design clean and professional. Make sure it's easy for readers to navigate and find information.
- It might take some iterations to finalize your newsletter design. That's perfectly normal.

STEP 4: Create and Publish Your Content (Ongoing)

- Keep your content informative and engaging. If you script your content, break up text with subheadings, bullet points, and images to make it easier to read.

- Once you're happy with your content, submit it for the newsletter. Make sure to proofread for any errors before submission.
- Plagiarism is a serious offense. Always attribute sources if you use someone else's work.

STEP 5: Promote the Newsletter (Ongoing)

- Share the newsletter on internal platforms like Slack or Microsoft Teams to reach a wider audience within your organization.
- Respond to feedback or questions about your content. This can help you build relationships and increase your visibility within the organization.
- Building an engaged readership takes time and patience.

STEP 6: Monitor Your Progress (Weekly/Monthly)

- Use feedback and engagement metrics to understand which types of content are most appreciated by your colleagues.
- Based on your analysis, adjust your strategy. Maybe your colleagues prefer certain topics over others, or they engage more with certain formats.
- Don't become discouraged by slow growth. It's better to have a small, engaged readership than a large, uninterested one.

Consistent contribution to the newsletter can position you as a thought leader, expand your internal network, and open up new opportunities like leadership roles within your organization. Keep refining your strategy based on your performance and goals. As you grow, consider exploring advanced tactics like guest contributions from industry experts or collaborations with other departments.

Newsletters stand as an instrument of power. Their true essence isn't in the sheer count of eyeballs they attract but rather in the wisdom disseminated and the relationships nurtured through their medium.

Section Summary

In essence, sharing our knowledge and experiences is a powerful tool for personal growth and understanding. As we break down our knowledge to teach others, we gain a deeper comprehension of our own wisdom. This act of vulnerability not only empowers others but also reinforces the insight within ourselves, creating a reciprocal relationship of growth and learning.

Sharing is likened to expanding the capacity of a suitcase—our hearts. With each shared story or piece of wisdom, we're not emptying our suitcase but rather making it bigger, creating more room for experiences, ideas, and emotions. However, it's important to remember that everyone has different capacities, and that's okay. Every time we share, we provide an opportunity for others to learn how to expand their own "suitcase."

Ultimately, sharing is a dance of comprehension and joy, each step expanding our capacity for the next. So, continue to share your wisdom, stories, and experiences. With every share, you step onto the dance floor of growth, expand your heart's suitcase, and become a bigger version of yourself.

CHAPTER SUMMARY

Life is like a suitcase. When you share your experiences and knowledge, it's like you're not emptying that suitcase but expanding it, making more room for new ideas, emotions, and experiences. It's a fascinating dance of joy and comprehension, and with each step, you grow into a larger version of yourself.

Here's what I want you to do: First, make sharing your wisdom a daily habit. Consistency is key; the more you do it, the more you grow. Second, remember that everyone has different capacities. Respect that, and grow at your own pace. Lastly, take the time to teach others. As you break down your knowledge to teach, you'll find that you comprehend your wisdom on a deeper level. So step onto that dance floor of growth and see how much your "suitcase" can expand!

CHAPTER 5

The Resolution: What's Next In Your Career Development

"The best way to predict your future is to create it."
—Abraham Lincoln

SECTION 1: REEVALUATE YOUR VALUES

As we journey through the landscape of life, we're inevitably shaped and molded by the experiences we encounter. This evolution isn't one-dimensional; it's a complex process that touches every facet of our being, including our values.

Our values aren't static or unchanging. Instead, they reflect the fluid nature of our identity—shifting and adapting as we traverse different stages of life. Authentic living demands that we align these values with our actions, ensuring our decisions mirror who we are at any moment.

Like historians critically examining sources, it helps to scrutinize our values regularly. This allows us to maintain consistency between our beliefs and actions, helping ward off feelings of discord or dissatisfaction. It's akin to sourcing information; it ensures our actions are rooted in our current beliefs, not outdated ones.

Examining our values also offers deeper self-understanding. It's a form of introspection that uncovers the motivations behind our behaviors. This insight isn't just personally enlightening—it's a valuable tool for navigating personal and professional realms.

Our values are more than mere beliefs. They serve as compass points guiding our life's purpose. When we reassess them, we verify that we're on the right path—our actions are meaningfully contributing to our broader life goals.

While our fundamental values may exhibit stability, it's critical—and indeed beneficial—to continually test and reevaluate them. As history is not a fixed narrative but an ongoing dialogue, so are our values in a dynamic conversation with ourselves as we grow and change.

Reevaluating Values: A Key Strategy for Sustainable Career Advancement

In the fast-paced world of tech startups, there was an exceptional engineering manager named Kate. Known for her technical proficiency and leadership skills, Kate navigated various challenges. However, she faced a different challenge as she climbed the career ladder.

Kate often found herself at the receiving end of crude jokes from her peers and even superiors. Initially, she brushed them off, considering it part of the startup culture. Over time, though, she began recognizing these "jokes" as subtle forms of workplace bullying.

This realization led Kate to introspect. She questioned why she had tolerated such behavior. Was it her desire to be accepted? Or was it an attempt to avoid conflict?

As Kate reflected on her values and priorities, she acknowledged they needed a reevaluation. She had put her professional growth above her self-respect. She permitted her boundaries to be crossed under the guise of humor.

Determined to bring about a change, Kate decided to redefine her boundaries. She resolved to stand up for herself, irrespective of the consequences. The path wasn't easy. While some colleagues respected her stance, others resisted the change.

Nevertheless, Kate remained steadfast. She communicated her concerns assertively, clarifying that disrespect would no longer be tolerated. She sought guidance from mentors and human resources, attended workshops on workplace etiquette, and learned strategies to handle such situations effectively.

Over time, Kate noticed a significant transformation. Her confidence grew, and her influence as a leader increased. The crude jokes gradually diminished, replaced by genuine respect. Her team's performance improved, and they began to see Kate as their manager and a symbol of respectful and dignified leadership.

Kate's journey is a powerful reminder to align our career choices with our core values. It underscores the need to stand up against workplace bullying, maintain self-respect, and set clear boundaries. It is a testament that earning a living should never come at the expense of self-respect and dignity.

Key takeaways:

1. The Necessity of Value Reevaluation: Kate's story demonstrates how periodic self-reflection and reevaluation of per-

sonal values can significantly change one's professional life. By recognizing the disconnect between her work experiences and core values, Kate could redefine her boundaries and stand up against disrespectful behavior. This highlights the importance of regular value reassessment in maintaining dignity and self-respect in the workplace.

2. Values as a Guiding Force: Kate's journey underscores that our values can guide professional decisions and actions. When she realized her tolerance for crude jokes compromised her self-respect, she realigned her efforts with her values. This shift improved her work environment and her influence as a leader, proving that a solid adherence to personal values can lead to respect and improved team dynamics.
3. The Role of Values in Personal and Professional Development: Kate's experience emphasizes the crucial role of a person's values in their personal and professional development. Despite resistance, her commitment to uphold her values led her to seek guidance, learn new strategies, and attend workshops. This dedication to aligning her work behavior with her values resulted in personal growth and professional success, highlighting the profound impact of value alignment on career progression.

Our values aren't static or unchanging. Instead, they reflect the fluid nature of our identity—shifting and adapting as we traverse different stages of life. Authentic living demands that we align these values with our actions, ensuring our decisions mirror who we are at any moment.

Examining our values also offers deeper self-understanding. It's a form of introspection that uncovers the motivations behind our behaviors. This insight isn't just personally enlightening—it's a valuable tool for navigating personal and professional realms.

Evolving Values: The Need for Periodic Reassessment in Personal Growth

Yes, it is important to reassess and test the values that we live by periodically. There are several reasons for this.

- Personal Growth: As individuals, we are constantly growing and evolving. Our experiences can significantly shape our perspectives and beliefs. What was important to us at one stage may no longer hold the same significance at another.
- Adaptability: Our priorities change as we go through different life phases. Our values must align with these changes to ensure we live authentically and make decisions that reflect our identity.
- Consistency: By regularly testing our values, we can ensure that we live in a way that is consistent with these values. This can help to prevent feelings of dissatisfaction or dissonance.
- Understanding: Testing our values helps us better understand ourselves. It can provide insight into why we behave the way we do and what motivates us. This self-awareness can be invaluable in both our personal and professional lives.
- Purpose: Our values often guide our sense of purpose. By reassessing them, we can ensure that we are on the right path and that our actions contribute to our overall life goals.

While our core values might remain relatively stable, it's healthy and beneficial to keep testing and reevaluating them as we grow and our circumstances change.

Navigating the Tech Terrain: A Guide to Aligning Your Values for Personal and Professional Growth

Navigating the tech industry can often feel like traversing a labyrinth. With its rapid pace, constant innovation, and diverse roles, it requires a unique blend of skills, values, and attitudes to thrive. The following advice has been carefully curated to provide general guidance for personal growth and career-specific insights to help you excel in your chosen role within this dynamic sector. Each piece of advice is designed to be universally applicable, but its manifestations may vary depending on your specific role and context.

In the subsequent sections, we will delve deeper into how these principles can be applied to various positions within a tech company, offering tangible examples and actionable steps. Prepare to embark on a journey of continuous learning, self-awareness, resilience, innovation, and ethical leadership as we explore the diverse tech industry landscape.

Continuous Learning and Growth

Embrace lifelong learning and constantly reassess your values. Stay open to new perspectives and experiences, which can significantly shape your beliefs and priorities.

Due to their dynamic nature, continuous learning and growth, especially in the context of reevaluating one's values, are crucial in the tech industry. This process helps individuals stay relevant, adaptable, and aligned with their personal and professional goals. Here's why:

1. **Adaptability:** The tech industry is characterized by rapid change and innovation. By constantly reassessing their values, tech professionals can adapt to new technologies, methodologies, and practices that align with their evolving values.

2. **Alignment with Company Culture:** Companies in the tech industry often have strong cultures defined by core values. Regularly reevaluating one's values can ensure alignment with the company's culture, increasing job satisfaction and productivity.
3. **Ethical Considerations:** The tech industry often grapples with ethical issues like data privacy and AI ethics. By reassessing their values, tech professionals can navigate these complex issues in a way that aligns with their personal beliefs and the expectations of society.

Here are three specific examples of individuals in the tech industry who might find value reevaluation particularly useful:

1. **Software Developers:** With the rise of DevOps and Agile methodologies, developers must constantly reassess their values to ensure they align with these iterative and collaborative approaches. For example, a developer prioritizing individual achievement might need to shift toward valuing collaboration and continuous improvement.
2. **Data Scientists:** As stewards of data, these professionals often face ethical dilemmas related to privacy and bias. A data scientist may need to reassess their values to prioritize fairness, transparency, and accountability.
3. **Tech Executives:** Leaders in tech companies must regularly reassess their values to guide their decision-making, set the company's vision, and build a strong culture. For example, an executive might reevaluate their values to prioritize diversity and inclusion in their organization.

To take actionable steps, these individuals could:

1. Conduct a personal values assessment to identify their current values.

2. Compare these values with their company's or desired role to identify gaps.
3. Develop a plan for personal growth that addresses these gaps, such as seeking training, mentorship, or new experiences.

Self-Awareness and Balance

Understand your strengths, weaknesses, and what motivates you. Seek a balance between personal life and work to maintain overall well-being.

Self-awareness and balance are critical in the tech industry because of the high-stress, fast-paced nature of the work. Understanding your strengths, weaknesses, and motivations can help you align your actions with your values, leading to increased satisfaction and productivity. Seeking a balance between personal life and work is essential for maintaining mental health and preventing burnout.

Here's why:

1. Job Performance: Self-awareness allows tech professionals to understand where their skills and values align, enabling them to perform at their best while feeling fulfilled.
2. Career Progression: Understanding your motivations can guide your career decisions, ensuring they align with your core values.
3. Work-Life Balance: The tech industry is notorious for demanding schedules. By prioritizing work-life balance, tech professionals can ensure living according to their values, improving well-being and job satisfaction.

Here are three specific examples of individuals in the tech industry who might benefit from self-awareness and balance:

1. UX Designers: These professionals must empathize with users to create intuitive designs. A UX designer might reassess their values to prioritize empathy and user-centric design, improving their ability to meet user needs.
2. Software Engineers: These professionals often work on complex projects under tight deadlines. By understanding their strengths and weaknesses, software engineers can better manage their workload and prevent burnout.
3. Tech Startup Founders: Founders often face high levels of stress and uncertainty. By seeking a balance between work and personal life, a founder can maintain their well-being while building their company.

To take actionable steps, these individuals could:

1. Use self-reflection exercises or psychometric tests to gain insights into their strengths, weaknesses, and motivations.
2. Identify areas of misalignment between their values and their current role or company.
3. Develop a plan to address these gaps, such as seeking new opportunities, setting boundaries for work-life balance, or pursuing personal development activities.

Belief and Resilience

Have faith in your abilities and don't fear failure. Life will throw challenges at you; it's not about avoiding them but how you bounce back from them that counts.

Belief in oneself and resilience are fundamental in the tech industry because of the high risk and rapid change inherent in the field. These qualities help individuals navigate challenges, learn from failures, and continually adapt their values to align with their experiences and goals.

Here's why:

1. Innovation and Risk-Taking: The tech industry thrives on innovation, often taking risks and confronting failure. By having faith in their abilities and resilience, tech professionals can embrace this aspect of their work and use it as a catalyst for growth.
2. Learning from Failure: Failure is a powerful learning tool in the tech industry. Resilience allows individuals to learn from their mistakes and improve while reassessing their values to better align with their learned experiences.
3. Adapting to Change: The tech industry is characterized by rapid change. Belief in one's abilities and resilience help individuals adapt to new technologies, methodologies, and practices that align with their evolving values.

Here are three examples of tech professionals who might benefit from these qualities:

1. Tech Entrepreneurs: These individuals often face numerous challenges and setbacks when starting a new venture. By believing in their abilities and demonstrating resilience, they can overcome these obstacles, learn from their experiences, and reassess their values to guide their entrepreneurial journey better.
2. Project Managers: These professionals often navigate complex projects with multiple stakeholders. Resilience can help them manage project setbacks effectively and reassess their values to prioritize effective communication, team collaboration, and problem-solving.

3. Cybersecurity Experts: These professionals often deal with high-pressure situations and constantly evolving threats. A strong belief in their abilities and resilience can help them handle stress and adapt to new challenges, while they also reassess their values to prioritize ongoing learning and ethical considerations.

To take actionable steps, these individuals could:

1. Reflect on past failures or challenges to identify learning opportunities and reassess their values.
2. Set personal and professional goals that align with their reassessed values and leverage their strengths.
3. Develop a resilience plan for handling future setbacks, such as seeking support, practicing stress management techniques, or pursuing further training.

Responsibility and Positivity

Understand that you are the architect of your own life. Your decisions and actions shape your destiny. Maintain optimism and a positive mindset to keep moving forward.

Responsibility and positivity are essential in the tech industry due to its dynamic and challenging nature. Taking responsibility for one's actions and maintaining a positive mindset allows individuals to navigate obstacles, learn from experiences, and continually reassess their values to align with their evolving goals and circumstances.

Here's why:

1. Navigating Challenges: The tech industry is filled with complex problems that require innovative solutions. Taking responsibility for one's decisions and maintaining a positive

mindset can help professionals overcome these challenges and turn them into opportunities for growth.

2. Learning and Growth: Continuous learning is a must in the rapidly changing tech world. Accepting responsibility for one's learning journey and remaining positive during the process can facilitate personal and professional development.
3. Team Dynamics: Tech professionals often work in teams. A positive attitude and a sense of responsibility can contribute to a productive and harmonious team dynamic.

Here are three examples of tech professionals who might benefit from these qualities:

1. Tech Entrepreneurs: Entrepreneurship is fraught with risks and challenges. By taking responsibility for their decisions and maintaining a positive mindset, entrepreneurs can navigate these obstacles, reassess their values based on their experiences, and persist in their ventures.
2. Software Developers: These professionals often have to solve complex coding problems. By taking responsibility for their work and maintaining a positive attitude, they can effectively tackle these challenges, constantly improve their skills, and reassess their values to prioritize quality and efficiency.
3. Product Managers: These professionals must balance the needs of customers, stakeholders, and their team. A sense of responsibility and a positive mindset can help them make effective decisions, manage conflicts, and reassess their values to prioritize customer satisfaction and product success.

To take actionable steps, these individuals could:

1. Reflect on past decisions, identify learning opportunities, and reassess their values based on their experiences.

2. Develop a positive affirmation or mantra that aligns with their reassessed values and helps maintain a positive mindset.
3. Make a plan to take responsibility for an area of personal or professional growth, such as learning a new skill or improving a relationship.

Hustle and Patience

Success requires hard work and time. Be patient, persistent, and willing to make the effort necessary to achieve your goals.

Hustle and patience are essential in the tech industry due to its competitive and fast-paced nature. Recognizing the need for consistent hard work (hustle) and the understanding that success takes time (patience) allows individuals to continually reassess their values, aligning them with their long-term goals and industry realities.

Here's why:

1. Achieving Goals: Success in the tech industry often requires complex problem-solving and innovative thinking, which demands consistent hard work. Patience is vital because these solutions may not come quickly or easily.
2. Adapting to Change: The tech industry is constantly evolving. Hustle ensures individuals keep up with these changes, while patience allows them to understand that mastery takes time.
3. Long-Term Success: In the face of setbacks or slow progress, hustle and patience can help tech professionals stay committed to their goals and maintain a long-term perspective.

Here are three examples of tech professionals who might benefit from these qualities:

1. Tech Entrepreneurs: These individuals often face a long road to success, filled with challenges and setbacks. By embracing hustle and patience, they can stay focused on their vision, putting in the necessary work and time to build a successful venture.
2. AI Specialists: These professionals must constantly learn and adapt as AI technology evolves. Hustle and patience can help them stay committed to their ongoing education and be patient with the pace of progress in this complex field.
3. App Developers: These professionals must continually update and improve their applications to meet user needs and stay competitive. Hustle ensures they do the necessary work, while patience helps them understand that building a successful app takes time.

To take actionable steps, these individuals could:

1. Set SMART goals that reflect their reassessed values and the need for hustle and patience.
2. Develop a daily routine or habit that embodies hustle, such as dedicating a set amount of time each day to learning a new skill or working on a project.
3. Practice mindfulness or other stress management techniques to cultivate patience and maintain focus over the long-term.

Purpose and Passion

Identify what you're genuinely passionate about and strive to make a career out of it. You spend much of your life working, so ensure you love it.

Purpose and passion are crucial in the tech industry due to its demanding nature. Aligning one's career with their passions ensures fulfillment in their work, even when facing challenges. This alignment also encourages individuals to continually reassess their values in light of their passions and purpose.

Here's why:

1. Job Satisfaction: Tech professionals passionate about their work are more likely to experience job satisfaction, which can lead to improved performance and productivity.
2. Motivation and Persistence: Passion for one's work fuels motivation and persistence, especially when encountering obstacles or setbacks.
3. Long-Term Commitment: A sense of purpose can foster a long-term commitment to one's career, encouraging ongoing learning and growth in the rapidly evolving tech industry.

Here are three examples of tech professionals who might benefit from aligning their careers with their passions:

1. Tech Entrepreneurs: Those passionate about innovation and solving problems can find fulfillment in starting their tech ventures. The cost of inaction could be missed opportunities for creating impactful solutions and personal growth.
2. Cybersecurity Experts: If one is passionate about protecting data and ensuring privacy, a career in cybersecurity could be fulfilling. The cost of inaction may lead to burnout or disinterest in a misaligned role.
3. Software Developers: Those who love coding and creating new applications can find immense satisfaction in software development. The cost of inaction could be a lack of motivation and reduced productivity in a job they don't love.

To take actionable steps, these individuals could:

1. Reflect and identify their passions and how they align with their current or potential tech industry roles.

2. Set career goals that align with their passions and reassess their values to ensure they support these goals.
3. Seek opportunities (like projects, courses, or roles) that align with their passions and contribute to their sense of purpose.

Innovation and Adaptability

In an ever-changing business landscape, stay innovative, adaptable, and focused. Strive to create something unique and always be ready to learn something new.

Innovation and adaptability are critical in the tech industry due to its rapidly changing nature. Staying innovative and adaptable allows individuals to keep pace with new technologies and methodologies, thus ensuring their relevance and effectiveness in their roles. Furthermore, this encourages individuals to continually reassess their values to ensure they align with an innovative and adaptable mindset.

Here's why:

1. Relevance: The tech industry evolves quickly, with new technologies and methodologies emerging regularly. To stay relevant, tech professionals must be innovative and adaptable, ready to learn and implement new trends.
2. Problem-Solving: Tech work often involves complex problem-solving. Innovation allows for unique solutions, while adaptability ensures these solutions remain effective in a changing landscape.
3. Career Progression: In a competitive industry like tech, innovation and adaptability can set individuals apart, leading to more opportunities and career progression.

Here are three examples of tech professionals who might benefit from these qualities:

1. Data Scientists: These professionals must stay up to date with the latest data analysis techniques and tools. Innovation and adaptability can help them leverage these to derive valuable insights effectively. The cost of inaction could be ineffectiveness in their role and missed career progression opportunities.
2. UX Designers: User preferences and technologies change rapidly. UX designers must be innovative to create engaging user experiences and adapt to new design principles and technologies. The cost of inaction could be creating irrelevant designs and falling behind in the industry.
3. Cloud Engineers: As cloud technologies evolve, these professionals must be adaptable to implement these changes and innovative to maximize the benefits of cloud computing. The cost of inaction could be inefficient systems and a lack of competitiveness.

To take actionable steps, these individuals could:

1. Engage in continuous learning, such as attending webinars, enrolling in courses, or reading up on the latest tech trends.
2. Practice innovative thinking, such as brainstorming sessions or participating in hackathons.
3. Regularly reassess and align their values with industry changes and personal career goals.

Value and Quality

Aim to provide value in whatever you do. Focus on doing one thing extraordinarily well. Your income is directly related to the value you provide to the marketplace.

Value and quality are fundamentally important in any industry, including tech. They are the key drivers determining business success, customer satisfaction, and personal fulfillment. Emphasizing value and quality encourages individuals to reassess their values, ensuring they align with these crucial aspects.

Here's why:

1. Business Success: Businesses prioritizing value and quality are more successful when meeting or exceeding customer expectations.
2. Customer Satisfaction: Delivering high-quality products or services with significant value increases customer satisfaction.
3. Personal Fulfillment: Focusing on providing value and quality can lead to personal fulfillment, as it often means positively impacting others.

Here are three examples of tech professionals who might benefit from these qualities:

1. Software Developers: By creating high-quality code that provides real user value, developers can ensure their work is impactful and fulfilling. The cost of inaction could be developing software that fails to meet user needs or market standards.
2. IT Project Managers: These professionals can aim to deliver projects that significantly value their organization while maintaining high-quality standards. The cost of inaction could be project failure or wasted resources.
3. Tech Consultants: By providing valuable, high-quality advice, consultants can help their clients succeed. The cost of inaction could be a loss of client trust and potential business opportunities.

To take actionable steps, these individuals could:

1. Regularly seek feedback to understand how their work provides value and where improvements can be made.

2. Pursue ongoing education and training to improve their skills and the quality of their work.
3. Reevaluate their values to ensure they align with a focus on value and quality.

Leadership and Communication

As a leader, inspire and motivate your team with empathetic and inclusive leadership. Effective communication is vital in any career.

Leadership and communication, particularly empathetic and inclusive leadership, are vital in the tech industry. These qualities foster a productive, diverse, and innovative work environment. Regularly reevaluating one's values to ensure they align with empathetic and inclusive leadership is crucial for tech leaders to be effective and impactful.

Here's why:

1. Team Cohesion: Empathetic and inclusive leadership fosters team cohesion, which leads to more effective collaboration and better results.
2. Workplace Diversity: Leaders who listen, learn, and understand others create an inclusive, diverse, and accessible workplace.
3. Innovation and Creativity: Empathy in leadership contributes to increased innovation, creativity, and job satisfaction.

Here are three examples of tech professionals who might benefit from these qualities:

1. Tech Team Leads: By adopting an empathetic and inclusive approach, team leads can create a supportive environment that fosters collaboration and innovation. The cost of inac-

tion could be team discord, reduced productivity, and high turnover rates.

2. IT Project Managers: These professionals can use empathetic and inclusive leadership to ensure all team members feel valued and heard, leading to successful project outcomes. The cost of inaction could be project failure or delays.
3. Tech Executives: By embodying these qualities, executives can set the tone for the entire organization, promoting a culture of inclusion and empathy. The cost of inaction could be a toxic work environment and reduced company performance.

To take actionable steps, these individuals could:

1. Engage in empathy training or workshops to develop their understanding and application of empathy.
2. Foster an inclusive culture by promoting diversity and ensuring all voices are heard.
3. Regularly reassess their leadership style and values to ensure they align with empathetic and inclusive principles.

Ethics, Integrity, and Inclusion

Uphold high ethical standards and integrity in your work. Embrace diversity and inclusion in the workplace, as different perspectives can lead to innovative solutions.

Ethics, integrity, and inclusion are pivotal in the tech industry, shaping not just individual actions but also companies' overall culture and reputation. Upholding high ethical standards, demonstrating integrity, and embracing diversity and inclusion can lead to innovative solutions and a positive work environment. Continually reevaluating these values is crucial to maintain trust, fairness, and innovation.

Here's why:

1. Trust: High ethical standards and integrity build trust among team members, stakeholders, and customers.
2. Fairness: Ethics and integrity ensure fairness, promoting a more inclusive and harmonious workplace.
3. Innovation: Diverse and inclusive teams bring different perspectives, leading to more innovative solutions.

Here are three examples of tech professionals who might benefit from these values:

1. Software Engineers: By upholding high ethical standards and integrity, they can create software that respects user privacy and security. The cost of inaction could be a loss of user trust and potential legal issues.
2. Data Scientists: These professionals deal with sensitive data and must maintain high ethical standards to respect privacy and prevent misuse. The cost of inaction could be unethical use of data and loss of trust.
3. Tech HR Managers: By fostering an inclusive and diverse workplace, HR managers can drive innovation and job satisfaction. The cost of inaction could be a homogenous workforce, reduced creativity, and potential discrimination issues.

To take actionable steps, these individuals could:

1. Attend training or workshops on ethics and integrity in the tech industry.
2. Actively promote and implement diversity and inclusion policies in their teams or organizations.
3. Regularly reassess their values to align with ethics, integrity, and inclusion.

ACTIVITY: Career Value Reevaluation and Roadmap Creation

This activity will help individuals in the tech industry select one to three pieces of advice from the guidance and advice list provided, dive deep into their meanings, customize them to fit their career paths, and chart a roadmap for habit formation and mental muscle development for continuous value reevaluation.

Materials Required:

1. Reference to the guidance and advice list in this section.
2. The notebook you've used while reading this book.
3. Pen or pencil.

Instructions:

1. **Selection (30 minutes):** Read the guidance and advice list carefully. Select one to three pieces of advice that resonate most with your career or aspirations.
2. **Deep Dive (1 hour):** For each selected piece of advice, write down what it means to you in the context of your career. Include personal examples or situations where this advice could be applied.
3. **Customization (1 hour):** Adapt each piece of advice to fit your career path. Define actionable steps to implement this advice in your daily work life.
4. **Roadmap Creation (2 hours):** Create a roadmap for the next six months based on your actionable steps. This should include clear goals, milestones, and timelines. Consider potential obstacles and how you might overcome them.

5. **Habit Formation (Ongoing):** Implement your roadmap. Regularly review and adjust it as needed. Make the process of value reevaluation a regular habit, such as a monthly or quarterly activity.

The initial stages of this activity (selection, deep dive, customization, and roadmap creation) can be completed in 1 day or spread over a week, depending on your schedule. The Habit Formation stage is ongoing.

Potential Outcomes:

Imagine gaining a crystal-clear understanding of your personal career values and how beautifully they align with your actions. Picture yourself crafting an actionable roadmap for career advancement, a path that's tailored to your unique aspirations and strengths. But don't stop there. Develop a regular habit of value reevaluation, turning it into a part of your routine. This practice will ensure continuous personal and professional growth. It's not just about reaching the pinnacle of success—it's about becoming the kind of person who is capable of achieving such success and maintaining it. So, get ready to embark on this journey of self-discovery and growth.

Takeaways:

Consider this: values are not etched in stone. They're dynamic, ever-evolving, much like we are. Thus, they demand our regular attention for reassessment and realignment with our actions. Now, how do you make this abstract advice palpable? Simple. With clear, actionable steps and a roadmap that turns these grand ideas into achievable tasks. But it doesn't end there. Regular reevaluation is the key that maintains this alignment between your values and actions, driving authentic career advancement. It's not just about climbing the ladder of success—it's about ensuring that every rung on that ladder resonates with who you are and what you believe in. So, gear up to take these steps and watch as your career transforms in ways you've only dreamed of.

Section Summary

As we close this chapter on reevaluating your values, it's essential to remember that the path to career advancement is not a straight line but a journey of continuous learning and growth. It demands self-awareness and balance between personal life and work.

Your resilience and belief in your abilities will help you face and overcome challenges. You are the architect of your life—your decisions and actions shape your destiny. Stay encouraged and ready to hustle while exercising patience, as success requires hard work and time.

Keep your purpose and passion at the forefront for those on a specific career path. In an ever-changing business landscape, staying innovative and adaptable is key. Aim to provide value and focus on quality in your endeavors.

As a leader, your communication skills and ability to inspire and motivate your team are crucial. Uphold high ethical standards and integrity in your work and embrace diversity and inclusion in the workplace.

Reevaluating your values is not a one-time activity but a regular practice that ensures your actions align with your beliefs and priorities. This alignment is the foundation of authentic career advancement and satisfaction.

Your values guide your career journey. Regularly reevaluating them ensures you stay true to your authentic self and enables you to navigate your path effectively. Invest time in this practice; you'll find it a powerful tool for personal and professional growth.

SECTION 2: REINVENT YOURSELF PERIODICALLY

The concept of reinvention is not new; it's a phenomenon observed across industries, wildlife, and human history. This section will explore why continuous reinvention is crucial for career development, particularly in the tech industry.

In the dynamic fashion world, reinvention is the key to staying relevant. Designers continually push the envelope, creating ingenious new styles and engaging with various influences from fashion's past. This constant cycle of reinvention has led to brands becoming favorites among young people, who are always looking for the latest trends.

Pioneers like Rewilder have taken reinvention a step further by addressing sustainability issues. By recovering waste materials, they undermine the legacy of waste perpetuated by the fast-fashion industry. Other entrepreneurs are branching off to create sustainable brands, focusing on organic or bio-based materials.

The impact on the ordinary person is significant. Consumers now can access various styles influenced by diverse cultures and eras. They also have the option to choose sustainable brands, aligning their fashion choices with their values.

In wildlife and human history, reinvention is a fundamental aspect of evolution. Species adapt and change to survive in changing environments. Similarly, human societies have reinvented themselves repeatedly, from agricultural to industrial revolutions, shaping our modern world.

In the technology industry, the pace of change is rapid. New technologies emerge, consumer preferences shift, and market dynamics fluctuate. To stay ahead, individuals working in tech need to continually reinvent themselves, learn new skills, adapt to changes, and innovate.

Let's dive deeper into the importance of reinvention in the tech industry. We'll explore how tech professionals can embrace this concept for career advancement and personal growth.

Adapting to Change: How Tech Professionals Can Learn from the Fashion Industry's Pioneers of Reinvention

The fashion industry is no stranger to reinvention. There have been several instances where designers, brands, or trends significantly altered the course of fashion history, impacting the industry, societal norms, and individual lifestyles.

Coco Chanel[42] played a pivotal role in liberating women from the constraints of the "corseted silhouette" and popularizing a sporty, casual chic as the feminine standard. She didn't invent women's pants, but she certainly made them fashionable at a time when women were taking on jobs traditionally held by men. Without her influence, the comfort and flexibility that characterizes much of women's fashion today might not exist. Like how Coco Chanel revolutionized women's fashion by popularizing pants, tech professionals can also embrace new roles and responsibilities that might seem unconventional or intimidating initially. For instance, a software engineer at a startup might be tasked with managing a team or coordinating with clients. While this might be outside their comfort zone initially, they can acquire a broader skillset and a holistic understanding of the business by embracing these new roles, making them more versatile and valuable in their career.

Yves Saint Laurent[43] was among the first designers to put women of color on the runway. He revolutionized the fashion industry by blurring gender lines, introducing the concept of luxury ready-to-wear, and making high fashion accessible to a broader audience. His efforts contributed to the diversity and inclusivity we strive for in today's fashion industry. In the tech industry, diversity is vital to fostering innovation and creativity. Akin to Yves Saint Laurent introducing women of color to the runway, tech companies can also strive to create diverse and inclusive workplaces. An anecdote from a tech giant reveals how they deliberately formed teams with members from different backgrounds, experiences, and perspectives. This resulted in innovative solutions and fostered a culture of inclusivity and respect.

The Industrial Revolution was a turning point in the fashion industry. It transformed clothing production, shifting from handmade items to mass production. This shift made fashion more accessible to the everyday person, paving the way for ready-to-wear trends. Just as the fashion industry adapted to the mass production of the Industrial Revolution, tech professionals must also adjust to technological advancements. A story shared by an IT professional recounts how they had to learn cloud computing to keep up with the shift from local servers to cloud-based systems. This willingness to learn and adapt to new technologies was crucial in helping them stay relevant and competitive.

Brands like Abercrombie & Fitch, Dior, and Lacoste have reinvented themselves over time. They adapted to changing trends and consumer preferences, ensuring continued relevance in the fashion industry. Tech professionals can take inspiration from brands that successfully reinvented themselves. For example, an executive at a tech company shared how they steered their company toward AI and machine learning when they realized their existing services were becoming obsolete. This strategic shift saved the company and propelled it to new heights.

These examples of reinvention in the fashion industry highlight the importance of adaptability and innovation in any industry, including tech. They underscore that to remain relevant and successful, one must be willing to evolve and reinvent oneself continually.

In the tech industry, just as in fashion, reinvention is not a choice but a necessity. The pace of change is rapid and relentless with new technologies, shifting consumer preferences, and emerging trends swiftly disrupting industries and rendering traditional business models obsolete.

Consider the story of Netflix, which started as a DVD-by-mail service before reinventing itself as an online streaming platform. This reinvention came from recognizing the potential of the Internet to revolutionize how we consume media. Today, Netflix is a global powerhouse in the entertainment industry, producing award-winning content viewed by millions worldwide.

Microsoft and Adobe are other notable examples. Both companies successfully repositioned themselves around the cloud, moving from packaged software to subscription-based services. Once focused on mobile devices, NVIDIA successfully shifted its focus to graphics and data centers.

The tech industry also has its share of pioneers who reinvented their fields. Companies like Relationship Hero, Hustle, and Honey applied innovative technologies to traditional sectors, creating new opportunities and setting new industry standards.

These examples highlight the importance of continuous learning, adaptability, and a willingness to embrace change for tech professionals. It's not just about upskilling; it's about reinventing yourself every three to five years to stay relevant and valuable in your career.

Reinvention in the tech industry often involves embracing new technologies, rethinking business models, and reconceiving processes and operations. As a tech professional, this might mean expanding your skill set, exploring new areas, or changing your career path. You're not limited to your initial or current job role in the tech industry. The possibilities for reinvention are endless.

Reinvention is a survival strategy in the tech industry. Embrace change, continually learn, and be willing to adapt. You'll secure your place in the industry and open doors to new opportunities and career growth.

Seeding Success: How Tech Professionals Can Grow New Opportunities Within Their Current Organization

Imagine you're standing at the edge of a vast field. This field represents your career, and each plant in it symbolizes a task or responsibility. Some plants are robust and tall, defining tasks that are important to your business, urgent, and contribute to your career growth. Others might be smaller, less significant tasks that don't align with these three vectors. Your challenge is to navigate this field and choose which plants to nurture. This is where the Goal Filter Tool comes into play.

This remarkable tool, as discussed in Chapter 3, is your compass. It's designed to help you evaluate each task based on its relevance to the business, urgency, and potential to fuel your career growth. Each task gets a score—either 1 or 0—for each vector. Only those tasks that score a 1 across all three vectors are worth your immediate attention. The rest? Delegate, postpone, or let them go. This is the first step toward a strategic transition.

But how do you know which tasks to include in your list? Start by analyzing your current job description. Add any missing items to your list. Then, think about your organization's key objectives that fall under your responsibility. Add those, too.

Now, let's dream a little. Think about the role you aspire to transition into. Research its job description and add those tasks to your list. But don't stop there. Extend your research to similar roles in competitive companies. What responsibilities do they encompass? Add those to your list, too.

This is not about working two jobs for the price of one. This is about creating a comprehensive list that reflects where you are and where you want to go.

Once you have this super-set of tasks, apply the Goal Filter Tool. Identify those tasks that score a 1 on all three vectors. This becomes your new task list, blending your current and aspiring roles.

Then, take this tangible list to your manager. Have a constructive discussion about your desire to transition, backed by clear evidence of what you can bring to the table. This is not just about advocating for yourself; it's about demonstrating how your transition can benefit the team, the business, the vision.

Use this list as a roadmap for skill development. Identify areas where you need to improve and seek out opportunities to grow.

This is a strategy born out of proactiveness. Paired with curiosity and the courage to ask questions, it's a powerful approach that reinforces your professionalism and signals your intent. It's an exciting journey of self-discovery and growth, a chance to reinvent yourself and create new opportunities. The effectiveness of this strategy largely depends on a few factors:

1. Understanding of the Goal Filter Tool: The tool is designed to help evaluate tasks based on their relevance to the business, urgency, and potential for career growth. It can significantly streamline your task list and focus your efforts on what truly matters if used correctly.
2. Proactiveness: This strategy requires you to take the initiative. You'll need to analyze your current job description, research

the role you aspire to transition into, and extend your research to similar positions in competitive companies.

3. Communication: A constructive discussion with your manager about your desire to transition is critical. You'll need to demonstrate how your transition can benefit the team and business and align with the larger vision.
4. Commitment to Skill Development: Identifying areas where you need to improve and seeking growth opportunities are essential. This might involve additional training or education, requiring time and commitment.
5. Attitude: Approaching the strategy with curiosity, courage to ask questions, and confidence in your abilities will significantly impact its success.

This strategy can be effective when executed properly. It promotes a proactive approach to career development within your current organization, encourages continuous learning, and fosters open communication with management. However, it's important to remember that individual results can vary based on variables such as the organizational culture, managerial support, and individual resilience and adaptability to change.

The Road to Renewal: A Guide for Tech Professionals Seeking Reinvention

The journey to reinvention for tech professionals involves a series of steps, each with its milestones and points of focus. As an illustration, let's consider the story of Alex, a software developer who felt a lack of passion for his work and decided to reinvent himself.

1. **Identify Your Driving Force:** Alex started exploring his interests outside of work and discovered a love for sustainable technologies. His driving force became clear—using his tech skills to impact the environment positively.

Milestone: Clearly articulating your driving force.

What to Pay Attention to: Your interests, what excites you, and what you naturally gravitate toward.

What to Look Out For: Don't mistake what you're good at for what you're passionate about. They're not always the same.

2. **Embrace Perseverance:** With the challenge of transitioning into a new field, Alex tried to learn new languages and understand different industries. Despite the difficulties, he persevered.

   Milestone: Accomplishing small goals that contribute to your more significant driving force.

   What to Pay Attention to: How you spend your time and energy. Make sure it aligns with your driving force.

   What to Look Out For: Avoid burnout. Hard work is essential, but so is rest.

3. **Continual Innovation:** Alex began experimenting with innovative ways to merge his tech skills with sustainable solutions. He developed a well-received app to track carbon footprints, encouraging him to innovate.

   Milestone: Successfully implementing a new idea or approach.

   What to Pay Attention to: Opportunities for innovation and improvement.

   What to Look Out For: Fear of failure. It can hinder innovation.

4. **Lifelong Learning:** To stay updated with the ever-evolving tech and sustainability sectors, Alex committed to constant learning, attending seminars, taking online courses, and reading about emerging trends.

Milestone: Completing a course, acquiring a new skill, or reading a set number of books per year.

What to Pay Attention to: Areas for growth and opportunities to learn.

What to Look Out For: Becoming complacent with your current knowledge and skills.

5. **Appreciate the Journey:** Despite feeling overwhelmed, Alex learned to enjoy the journey of growth and discovery, celebrate each small victory, and learn from setbacks.

   Milestone: Recognizing and celebrating personal growth and development, even if you haven't reached your end goal yet.

   What to Pay Attention to: Your growth journey and the lessons learned.

   What to Look Out For: Forgetting to appreciate small wins and progress.

6. **Act on Impulse:** At a tech conference, Alex had an idea for a new app feature. Instead of waiting, he acted swiftly on his impulse, which led to many successful features in his app.

   Milestone: Moving swiftly on an idea or inspiration as soon as it surfaces.

   What to Pay Attention to: Moments of sudden inspiration or motivation.

   What to Look Out For: Delayed action. Don't let a promising idea slip away because you didn't act on it promptly.

7. **Create Your Unique Path:** Realizing his journey was unique, Alex drew inspiration from other tech professionals but forged his path, staying true to his passion and using his skills authentically.

Milestone: Crafting a personal roadmap for reinvention that feels authentic and exciting to you.

What to Pay Attention to: What works for you and what doesn't. Adjust your strategy as needed.

What to Look Out For: Trying to follow someone else's path too closely. Forge your own.

Alex's story serves as an inspiration for all tech professionals seeking to reinvent themselves. It demonstrates the power of a precise driving force, perseverance, innovation, lifelong learning, appreciation for the journey, prompt action on impulses, and the courage to create your path. This strategy can be used as an outline for future descriptive activity, guiding users through each step with exercises and prompts.

8. **Develop the Edge:** Alex realized that to stay ahead and continuously reinvent himself, he needed to develop a competitive edge. He began to specialize in a niche area of sustainable technology, becoming an expert in his field.

   Milestone: Becoming recognized as an expert in a specific area or field.

   What to Pay Attention to: Emerging trends and niche areas within your field that align with your interests and skills.

   What to Look Out For: Becoming too broad in your focus. Specialization can help you stand out and provide more profound value.

   With this additional step, tech professionals reinvent themselves and ensure they maintain their relevance and competitive advantage in their chosen field. This approach keeps their skills and knowledge fresh, making it easier for them to adapt to changes and seize new opportunities as they arise.

ACTIVITY: Roadmap to Long-Term Career Reinvention Activity

This activity aims to guide tech professionals through the process of career reinvention. The goal is to complete these steps once in the immediate future and then revisit them every 12 to 18 months for course adjustment. It's a cyclical process that requires discipline, reflection, and continuous learning.

Materials Required:

- A notebook (the one you've been using throughout this course)
- Pen or pencil
- Digital tools for reminders and tracking (such as Google Calendar, Trello, or Asana)
- Access to online resources for research and learning

Instructions:

1. **Identify Your Driving Force:** Reflect on your interests and passions. Document them in your notebook. Use mind mapping or brainstorming software like MindMeister or Coggle to visualize your thoughts. *Timeline: 1 week*
2. **Embrace Perseverance:** Set small goals that align with your driving force. Use project management tools like Trello or Asana to track your progress. Regularly review and adjust your goals. *Timeline: Ongoing*
3. **Continual Innovation:** Identify opportunities for improvement in your field. Implement a new idea or approach. Use digital platforms like Google Docs or Evernote to document your thoughts and reflections. *Timeline: 1-2 months*

4. **Lifelong Learning:** Commit to constant learning. Use Coursera, Udemy, or LinkedIn Learning to access online courses and keep current with emerging trends. *Timeline: Ongoing*
5. **Appreciate the Journey:** Document your journey and lessons learned. Celebrate personal growth. Use journaling apps like Day One or Journey for regular reflection. *Timeline: Ongoing*
6. **Act on Impulse:** When inspiration strikes, act promptly. Use note-taking apps like Evernote or Google to jot down ideas as they come to you. *Timeline: Ongoing*
7. **Create Your Unique Path:** Reflect on what works and doesn't. Adjust your strategy as needed. Use your notebook to document these reflections. *Timeline: Ongoing*
8. **Develop the Edge:** Identify emerging trends and niche areas within your field. Begin to specialize in a niche area. Use tools like Feedly or Pocket to stay updated with relevant industry news. *Timeline: 6-12 months*

Outcomes:

You will embark on a transformative journey of career reinvention that will reshape both you and your professional trajectory. You will begin by identifying your driving forces, digging deep to discover the passions and motivations that will fuel your ambition and steer your choices. This will go beyond merely recognizing your strengths and weaknesses.

Next, you will develop habits of perseverance. You will foster resilience and tenacity with daily habits that will propel you forward, even in the face of adversity. Then, you will learn to implement innovative ideas. You won't just create something new; you'll challenge the status quo, push boundaries, and bring a fresh perspective to your field.

A crucial part of this journey will be committing to lifelong learning. In our ever-evolving world, continuous learning is not just beneficial, but essential. It will help you stay relevant, broaden your horizons, and unlock fresh opportunities.

Along the way, you will learn to appreciate your growth journey. You will recognize and celebrate your progress—every step, stumble, and triumph will form a unique chapter in your story. Be ready to act swiftly on impulses and ideas when inspiration strikes. Quick action can lead to breakthroughs and open doors to unexpected opportunities.

As you continue on this path, you will shape a career that mirrors your individuality, aligns with your driving forces, and sets you up for success. Finally, you will strive to develop a competitive edge. You will aim to be the best version of yourself by honing your skills, deepening your knowledge, and carving out a unique niche in your field.

Remember, this journey won't be solely about reaching a destination. It will be about personal and professional growth, achieving career fulfillment, and evolving into the best possible version of yourself. By following this roadmap, tech professionals can successfully reinvent their careers. The goal is not only to complete these steps but also to internalize them as part of a continuous process of growth and reinvention. Remember to revisit this activity every 12 to 18 months to adjust and keep your career journey aligned with your passions and the evolving industry landscape.

Section Summary

We delved into the journey of reinvention for tech professionals. We defined this process as a series of steps requiring self-reflection and action. First, we identified the importance of recognizing your driving force, which requires an exploration of your interests and passions. This forms the foundation of your career reinvention journey. Perseverance was recognized as a critical attribute, emphasizing the necessity of setting and accomplishing small goals that align with your larger career objectives.

We discussed the vital role of continual innovation in the tech industry, encouraging the embrace of opportunities for improvement and the courage to overcome the fear of failure. The concept of lifelong learning was introduced to stay updated with the rapidly evolving tech sector. It involves seeking out areas for growth and avoiding complacency.

We stressed the importance of appreciating the journey, celebrating personal growth, and learning from setbacks. This contributes to a healthy mindset and promotes resilience in facing challenges. The value of acting swiftly on impulses and ideas was also highlighted. This proactive approach often leads to success and encourages tech professionals to create unique paths, instead of strictly adhering to others.

Lastly, we emphasized the need to develop a competitive edge by specializing in niche areas within the field. This strategic move can help maintain relevance and provide a competitive advantage in the ever-changing tech industry. We used Alex's story as an inspirational example throughout this process. His journey serves as a guide for all tech professionals seeking to reinvent themselves. We provided a step-by-step activity using traditional and new-age tools regarding practical application. The goal is to initially complete these steps and revisit them every 12 to 18 months for necessary course adjustments.

The journey to reinvention for tech professionals is an ongoing, cyclic process that requires self-awareness, perseverance, continuous learning, and strategic planning. Each individual's journey will be unique, and success lies in authenticity and the ability to adapt to the evolving landscape of the tech industry.

SECTION 3: BECOME A LIFELONG STUDENT

As we stand on the precipice of our journey's conclusion, let's dive into what it means to be a lifelong student. This concept is not confined to accumulating skills or knowledge; it embodies an attitude of perpetual growth and adaptability.

In our age, where information is but a mouse click away, some may question the importance of retaining the mindset of lifelong learning. However, the relevance of this principle is more significant than ever.

The swift tide of technological evolution demands an unwavering commitment to ongoing education. Staying abreast of these shifts enables tech professionals to remain competitive and relevant. Moreover, this commitment acts as a catalyst for innovation. The more you learn, the more you can apply your new skills and knowledge to your work, paving the way for creative solutions and advancements.

It may seem challenging to cultivate a habit of continuous learning amidst a hectic schedule. Nevertheless, with strategic planning and discipline, it becomes achievable. Begin by setting clear learning objectives and incorporating learning into your daily regimen. Harness the power of online learning platforms, engage with professional networks, and practice reflective learning regularly.

The rewards of being a lifelong student extend far beyond career advancement. It prepares you with the necessary skills and knowledge to excel in your current role and primes you for future opportunities. It bolsters personal growth, enhances your value as a team member, and establishes you as a thought leader in your field.

In closing, *Hurry Up Slowly* invites you to embrace the journey of continuous learning. It's not merely about professional development but personal growth, adaptability, and preserving a sense of curiosity and wonder in a rapidly evolving world.

But let's not sugarcoat reality. There are reasons some tech professionals might halt their learning journey:

- Lack of Training Opportunities: Not all companies provide adequate training or learning opportunities.
- Fear of Learning New Tech Skills: Learning new tech skills can be intimidating for some.
- Work-Life Balance: Finding time for continuous learning can be a struggle in tech jobs.
- Inadequate Recognition: Motivation can wane if efforts to learn and grow go unnoticed.
- Perceived Threat of Digitalization: Some see technology more as a threat than an opportunity, discouraging them from acquiring new tech skills.
- Overwhelm Due to Rapid Technological Changes: The pace of technological change can be overwhelming.

Despite these challenges, the importance of continuous learning in the tech industry is undeniable. As we bid adieu, remember that the strategies shared in this book are not one-time actions. They demand consistent application and refinement. So, as you turn the final pages, view it not as the end but the commencement of a new chapter in your career development. Embrace the lessons from *Hurry Up Slowly* and observe how they transform not just your career but your life.

The Journey of a Perpetual Learner

A. Decoding the Lifelong Learning Philosophy

Lifelong learning is more than a concept—it's a way of life that flourishes on the relentless pursuit of knowledge for both personal and professional growth. It's about fostering an open mind, a curious spirit, and an unyielding thirst for self-improvement.

In the fast-paced realm of technology, lifelong learning is not just beneficial—it's essential. It's not merely about staying up to date with new technologies or best practices; it's about grasping the foundational principles driving these advancements. This robust understanding equips you, the tech professional, with the agility to adapt swiftly and efficiently, no matter what the future brings.

Remember the lessons from Chapter 1 about the power of knowledge? Lifelong learning is precisely that—leveraging knowledge as a tool for continuous growth and success.

B. Fostering an Attitude of Continuous Growth and Adaptability

Nurturing a lifelong learning mindset requires embracing change as an ally and perceiving challenges as stepping stones toward growth. Here's how:

- Ignite Curiosity: Spark a sense of wonder about the world. Seek answers and strive to understand the "why" behind things, just like we discussed in Chapter 3 about aiming at your North Star.
- Embrace a Growth Mindset: Trust in your ability to learn and evolve. Remember the lessons from Chapter 4 about consistency and habit-building? Apply them here. View failures not as stumbling blocks but as opportunities for growth.
- Reflect Often: Take time to introspect and learn from your experiences. This practice enhances self-awareness and fosters personal growth, similar to the reevaluation of values we explored earlier in this chapter.
- Welcome New Experiences: Embrace novel experiences with an open heart and mind. They offer fresh perspectives and learning opportunities.

- Strike a Balance: Remember to maintain equilibrium while chasing growth. Rest and rejuvenation are crucial for a mind ready to learn and grow.

For tech professionals, adopting a lifelong learning mindset is paramount. The technological landscape is in constant flux, and those who can adapt will flourish. This mindset allows you to navigate uncertainties, overcome challenges, and seize opportunities, fueling personal and professional growth.

The pursuit of lifelong learning is not a destination but an ongoing journey. It's about celebrating progress and persistently aiming for growth and excellence. So, embark on this enlightening journey with an open heart and mind, and unlock the infinite potential that lies within you.

The Unseen Driver in the Tech World: Continuous Learning

Imagine walking into your tech firm on a Monday morning. The usual hum of servers greets you, as well as the chatter of your colleagues discussing the latest tech trends and the aroma of freshly brewed coffee. Yet, amidst this familiarity, you sense an underlying current of change.

Your company just adopted a new technology that promises to revolutionize your products. As exciting as this is, it also means you need to learn a new set of skills—fast. How do you adapt? This is where the essence of continuous learning comes into play.

Continuous learning isn't about mindlessly accumulating knowledge. It's about fostering a deep-rooted curiosity that pushes you to question, explore, and innovate. Let me tell you about a colleague of mine. Let's call him John. John always asked questions about our products, processes, and the "why" behind our strategies. Some found it annoying, but his relentless curiosity led him to develop a feature that significantly improved our product's user experience. Now, that's the power of continuous learning!

In the ever-evolving tech landscape, continuous learning is about staying ahead of the curve, understanding the latest trends, and leveraging them to make a difference. Think about the most successful people in your company. Aren't they the ones who are always learning and constantly growing?

Continuous learning also plays a profound role in career progression. It equips you with the skills and mindset needed to tackle challenges, seize opportunities, and climb the success ladder. Remember Sarah, the intern who joined us last year? Within months, she learned multiple programming languages, embraced every challenge, and now she's leading her team. That's the impact of continuous learning on career progression.

However, continuous learning isn't a destination; it's a journey that requires dedication, resilience, and a love for learning. It's about embracing the process, celebrating small victories, and never stopping in the pursuit of excellence.

Learning isn't confined to books or classrooms. Every conversation is an opportunity to learn. So, keep your mind open, stay curious, and never stop asking questions. Remember to reflect. Reflection not only consolidates your learning but also promotes self-awareness and personal growth.

While chasing growth, don't forget to maintain balance. Take time for rest and relaxation. A well-rested mind is more receptive to learning and growth.

Overcoming Challenges to Continuous Learning: A Six-Step Activity Series

ACTIVITY 1: Create Your Own Learning Opportunities

This activity will help you overcome the challenge of a lack of training opportunities.

Instructions:

1. Identify your learning goals.
2. Research online platforms like Coursera, Udemy, or Khan Academy for courses that align with these goals.
3. Allocate specific time slots for learning and stick to them.

Start today and continue on an ongoing basis.

Potential outcomes include an enhanced skillset, self-reliance in learning, and increased professional value. Learning is not confined to formal training programs. With the right resources and dedication, you can take charge of your own learning journey.

ACTIVITY 2: Embrace Tech Skills

This activity will help you overcome the fear of learning new tech skills.

Instructions:

1. Identify a tech skill you want to learn.
2. Break down the learning process into manageable steps.
3. Join online forums or communities related to this skill for support.

Start today and continue as per the complexity of the skill chosen.

Potential outcomes include increased confidence, adaptability, and a valuable addition to your skillset. Every expert was once a beginner. With patience and perseverance, you can conquer your fear and master new tech skills.

ACTIVITY 3: Balance Work, Life, and Learning

This activity will help you maintain a work-life balance while incorporating continuous learning.

Instructions:

1. Define your daily routine.
2. Identify gaps or periods of downtime in which you can incorporate learning.
3. Include rest periods in your schedule.

Start today and adjust over the next few weeks as needed.

Potential outcomes include improved time management, reduced stress, and consistent learning without burnout. Efficient time management can help you balance professional, personal, and learning commitments effectively.

ACTIVITY 4: Celebrate Personal Growth

This activity will help you overcome the challenge of inadequate recognition.

Instructions:

1. Set small, measurable learning goals.

2. Celebrate each time you achieve one of these goals.
3. Discuss your learning efforts and achievements with your supervisor or HR if needed.

Start today and continue on an ongoing basis.

Potential outcomes include increased self-motivation, recognition of your efforts, and a sense of accomplishment. External recognition is rewarding, but the real value of learning lies in personal growth and the skills you acquire.

ACTIVITY 5: Embrace Digitalization

This activity will help you overcome the perceived threat of digitalization.

Instructions:

1. Identify a digital tool or technology relevant to your work.
2. Learn about this tool and how it can benefit your work.
3. Implement this tool in your work processes.

Start today and continue as per the complexity of the tool chosen.

Potential outcomes include enhanced digital literacy, increased efficiency, and staying relevant in the digital age. Instead of viewing digitalization as a threat, see it as an opportunity to enhance your skills and stay ahead of the curve.

ACTIVITY 6: Navigate Rapid Technological Changes

This activity will help you overcome overwhelm due to rapid technological changes.

Instructions:

1. Prioritize tech trends based on their relevance to your role or interests.
2. Stay updated on these trends through blogs, podcasts, or newsletters.
3. Learn one aspect at a time, focusing on understanding rather than memorization.

Start today and continue on an ongoing basis.

Potential outcomes include increased adaptability, staying updated with relevant trends, and reduced overwhelm. You don't have to learn everything at once. Prioritizing and taking one step at a time can help you navigate rapid technological changes effectively.

Continuous learning is a journey filled with challenges, but each challenge offers an opportunity for growth. Embrace these challenges, adapt, learn, and grow. After all, it's the challenges that make the journey worthwhile.

"The illiterate of the 21st century will not be those who cannot read and write, but those who cannot learn, unlearn, and relearn."
—Alvin Toffler

ACTIVITY: Fuel Your Tech Career with Continuous Learning

This activity is to cultivate a robust, adaptable, and forward-thinking learning habit that capitalizes on every opportunity, embraces technological change, and drives personal and professional growth in the tech industry.

Instructions:

1. **Targeted Tech Goals:** Set ambitious learning objectives tailored to your tech career. For example, if you're a software developer interested in AI, your goal might be to master Python and TensorFlow within 6 months.
2. **Everyday Tech Learning:** Incorporate tech-related learning into your daily routine. Listen to podcasts like "Coding Blocks" during your commute, read books like *Clean Code* by Robert C. Martin during breaks, or watch coding tutorials on YouTube in your free time.
3. **Leveraging Digital Platforms:** Make full use of digital platforms for learning. Engage in tech forums like Stack Overflow, enroll in online courses on Coursera or Udemy, and explore open-source projects on GitHub.
4. **Networking in the Tech World:** Connect with like-minded professionals through platforms like LinkedIn, join tech-related groups on Facebook, participate in virtual tech conferences, and attend local meetups (like those found on Meetup.com).
5. **Reflective Learning Applied to Tech:** After each learning session, consider how to apply your new knowledge to your current projects. For instance, how could the machine learning algorithms you learned optimize your company's data processing system?
6. **Tech-Focused Journaling:** Maintain a dedicated journal for your tech learning journey. Document new concepts learned, problems solved, and ideas for future projects. This journal can be a reference for job interviews, performance reviews, and personal audits of your learning progress.

Begin immediately, making this an ongoing commitment. Conduct monthly reviews to track progress and adjust goals as necessary.

From this activity, potential outcomes include a resilient learning habit tailored to the tech industry, a significant increase in technical knowledge and skills, enhanced problem-solving abilities, a detailed record of personal growth, and a persistent mindset of continuous improvement.

Continuous learning is crucial in the fast-paced tech industry. By turning challenges into opportunities, adapting to technological advancements, and constantly striving for excellence, you can stay ahead of the curve and boost your career trajectory.

SECTION SUMMARY: THE GRAND ADVENTURE OF LIFELONG LEARNING IN THE TECH INDUSTRY

Throughout this section, we highlighted how vital lifelong learning is in the tech industry. In a world where you can only count on change, continuous learning is the lighthouse guiding you through the stormy seas of technological evolution. My path serves as living proof of this.

As the first engineer in my family, I started my academic voyage in aerospace engineering. However, life had other plans, and I switched majors twice, from mechanical to electrical engineering. This was my introduction to the exhilarating dance of adaptability and the ceaseless pursuit of knowledge.

The first steps of my career were taken as an analog designer, an experience that propelled me to pursue a master's degree. The journey led me to California's golden shores, where I immersed myself in the art of designing analog circuits at the forefront of technology. Every shift demanded new skills, a bottomless well of knowledge, and an insatiable hunger for learning.

The adventure didn't stop there. I ventured into the realm of the Internet of Things, or IoT, applications engineering, engaging with customers around the globe and participating in conferences and events. Later, I dove into the world of high-speed transceivers, each step broadening my horizons, testing my limits, and reinforcing the power of lifelong learning.

The zenith of my corporate journey saw me leading a global team of technical engineers, training customer engineers on all facets of FPGAs (Field Programmable Gate Arrays)—hardware, software, and platform. Having navigated numerous stakeholder engagements and ensuring my team could flourish without me, I set sail on a new expedition: entrepreneurship.

Launching Grounow was akin to diving into the deep end of business. I grappled with accounting, website development, content creation, social media engagement, lead generation, marketing, sales, and even development of a mobile app. Each day served up a lesson. Each challenge morphed into a mentor.

I did all this while juggling family responsibilities—walking my children to school daily, maintaining multiple revenue streams, sharing household duties, and growing alongside my supportive spouse, who makes me feel like the most fortunate person alive.

As we wrap up this book, I implore you not to view it as an end but as the launchpad for your unique journey. The strategies and insights shared here are your arsenal for success, waiting to be sharpened and adapted to your path.

The tech world is a vast, ever-shifting universe. With lifelong learning as your guiding star, you are more than equipped to chart your course. Your next chapter is about to unfold.

CHAPTER 6

Letters For You From Your Tribe

AN UNSUSPECTING GIFT

As the curtain draws back for our epic journey, I present Chapter 6—a treasure trove of wisdom from fellow tech professionals. They have generously penned heartfelt letters, each a glowing ember of advice and guidance, illuminating your path in this grand narrative.

Picture it as a scene where you, our main character, receive an unexpected package. The seal is broken and out pours a cascade of letters, each one a unique voice echoing through the annals of the tech world. They speak of trials and triumphs, lessons learned, and wisdom earned, intending to enrich your experience with this book.

Their words will guide you through the thickets of challenge and the forests of opportunity. They'll shed light on the pitfalls to avoid and the ladders to climb, making your journey a little less daunting and much more enlightening.

Including these letters in this chapter is akin to a cameo appearance by renowned stars in our movie. Their presence adds depth, their experience provides insights, and their wisdom paves the way for your success.

And I must pause here to express my profound gratitude and honor. Releasing these letters to share this collective wisdom with you fills me with awe and humility. These aren't just letters; they are lifelines extended by those who have walked the path before you.

So, dear reader, as you read Chapter 6, know that you're about to unwrap a gift from the tech community. Each letter is a testament to their generosity and commitment to your success. Read them, cherish them, and let their words guide you as you continue your extraordinary journey with *Hurry Up Slowly*.

Letter 1

Dear fellow tech professionals,

In the fast-paced world of technology, it's crucial to navigate your career with purpose and resilience. With that said, here are three pieces of advice I've gathered through my career journey and career transition:

1. *Embrace Lifelong Learning and Stay Curious: The tech industry evolves rapidly and never seems to stop. It's essential to stay curious about the tech industry and never stop learning. While you don't need to dedicate hours every day to learning new topics, setting aside time each week or a few times a month to explore new skills is essential. If you fail to do this, you risk stagnating in your career.*
2. *Investing in your learning through online courses, books, or conferences is an investment in your future success. Being curious about the vast world of tech fuels new inno-*

vations in your career. Keep asking questions, challenging assumptions, and exploring new ideas. The tech industry rewards those who dare to be curious and think outside the box.

3. *Seek Mentorship: Don't hesitate to seek guidance from experienced professionals. A mentor can provide valuable insights, help you set goals, and navigate the complexities of the industry. Maya and I have found mentors to be invaluable in our careers.*

Fail Forward: Don't fear failure; embrace it as a stepping stone to success. Every setback is a lesson in disguise. Learn from your mistakes and keep moving forward. I have faced many challenging points in my career transition. There were definitely times when things were so difficult that I considered abandoning my transition and returning to what I was comfortable with. However, thanks to the advice and support of my mentors, I was able to push through these tough moments and make it to the other side.

Remember, your career in technology is a journey filled with opportunities and challenges. Keep learning and pushing your boundaries. Your dedication will undoubtedly lead you to great heights.

Best regards,

Aviv Weinstein | ML Infrastucture Software Engineer

Letter 2

Advice for Technical Professionals

1. *Efficient Time Management:*

   While working diligently is important, it's equally crucial to work smartly. In the world of technology, time is often of the essence. Use your time wisely by optimizing your workflow. Leverage tools and automation for routine and repetitive tasks. This not only saves time but also minimizes errors, allowing you to focus on more critical aspects of your work.

2. *Effective Communication:*

   Maintain open and proactive communication with your colleagues, managers, and even cross-functional teams. A strong network and clear lines of communication ensure you have a comprehensive understanding of projects, stay updated on crucial information, and receive valuable support. Beyond work-related benefits, this communication can lead to career inspiration, creative ideas, and even life lessons, as diverse perspectives enrich your professional and personal growth.

3. *Prioritize Life-Work Balance:*

   While intense dedication to work might seem like the path to success, it's important to strike a balance between your professional and personal life. Overworking can take a toll on your health. Realize that maintaining good health is not only a part of a successful career but also essential for a fulfilling life.

Reevaluate your work hours and habits. Your 20s may have allowed for long workdays, but as you enter your 30s, it becomes evident that excessive work can negatively impact your well-be-

ing. To ensure a long and successful career, make life-work balance a priority.

Remember that work is just one facet of your life. It's essential to allocate time for family, friends, and, most importantly, for yourself. A well-rounded life enhances your overall well-being, making you not only a better professional but also a happier individual.

Yi-hui Hsieh | Former IC design engineer

Letter 3

Dear Technical Managers,

As leaders in technology, we have a multifaceted responsibility that extends beyond the mere orchestration of tasks and projects. Our role encompasses the logical coordination of technical endeavors and the crucial consideration of the emotional landscape that drives our teams. Emotional engagement isn't necessarily top of mind, but it needs to be.

The rewards of leading technical teams are tremendous, offering the satisfaction of steering cutting-edge projects that push the boundaries of innovation. However, it's equally imperative to acknowledge the challenges of this leadership journey—foremost among them being the art of balance. Balancing the allure of pioneering, challenging projects with the necessity of tending to less-glamorous tasks is a continual tightrope walk.

Engaging technical talents in projects that stimulate and challenge their intellect is essential for their sustained motivation and growth. However, the reality of our roles dictates that we must also guide our teams through less-exhilarating projects—those that are essential but might need more of the spark of the bleeding edge of innovation. These 'trophy' projects are often exciting and seen by our teams as critical for their long-term recognition and career growth as technical professionals. Balancing these diverse demands requires finesse, understanding, and a deep appreciation of the bigger picture.

Continual innovation and professional growth are the lifeblood of technical excellence. Encouraging an environment that fosters experimentation and welcomes failure is critical to individual and organizational success and cultivates a learning culture. Our responsibility is to drive projects forward and to nurture an ecosystem that thrives on pushing boundaries, exploring uncharted territories, celebrating the most exciting successes, and acknowledging failures.

In this journey, we cannot overstate the role of mentors. As leaders, it's our responsibility to provide guidance and support, but it's equally critical to encourage mentorship within our teams. Mentoring not only aids in developing junior talents but also enriches the mentor. The reward lies in the technical prowess imparted and the profound satisfaction of nurturing and witnessing growth in others.

Ultimately, the true essence of leading technical teams lies in finding a delicate balance between logical and emotional aspects, pushing boundaries, and tending to the essentials, fostering innovation, and ensuring stability. As we navigate these challenges, let us remain resolute in our commitment to managing projects and inspiring, mentoring, and creating an environment where every technical mind flourishes. What a group of inspired, motivated, and diverse technical organizations can accomplish is truly unique.

Jack Dunnigan | Business and Sales Executive

Letter 4

Dear Reader,

Do you ever wonder what can accelerate your career other than having some technical skills? Well, I guess you would wonder because that would be why you are reading this book.

Let me tell you, coming out of college, my GPA was not stellar. At a career fair, I would show my résumé with my GPA and immediately get turned away. That was until I decided to address the GPA and what I could do that my GPA does not indicate. That was how my post-college career and networking started.

I would like to share with you a few of the unwritten and occasionally unspoken tips that helped me show up for myself in my nine-ish years since college.

Networking

The lady at the career fair I spoke to connected me to a network within her company. She handed my résumé to a direct report of hers, and eventually, I started an internship after graduation. Since that internship, I am now in my fourth role in the company, all of them having people that connected me to my next role.

Find the Coaches, Mentors & Sponsors

Your manager(s) is not/are not the only one(s) you can learn from. As you network, you will find people that can walk you through step by step, guide you, or speak on your behalf. Most of the time you will know who these people are. Sometimes, you may not even know who is watching your work. Keep connections to those that can be your coach, mentor, AND sponsor. Constantly stay in touch with them regardless of whether they work for your current company or not—you never know what opportunities there might be.

Ask the Questions and Some Organization Skills

So, you may have heard people state, "There's never a dumb question." Well, it's true! Asking questions provides clarity on expectations and the scope of work. As I was a new college grad, I wish I had someone to instill in me then what kind of questions to ask to better understand the scope of a project. My mentor and I realized at the end of a project when I was presenting it that either I could have asked more questions, or they could have provided me with more detailed explanations as to what I was doing and why. I ask my leadership today a lot more questions than I did as a new college grad as I am doing projects.

Organizational skills might be an often overlooked soft, and very transferable, skill. I am not talking just about having a tidy desk. I mean having some idea of structure and organization helps in your favor. This gives you the opportunities to lead cross-functional teams and show results in a timely manner. Recall I said you never know who is watching your work? Being even slightly organized will turn the heads of your leadership and colleagues.

Okay, These May Be Intuitive and Unoriginal. So What?

In 2023, I was given the opportunity to mentor a couple of younger employees, and I provided them with all the tips I mentioned above. One thing I added to my repertoire of tips was the importance of wellness.

I have been on a wellness journey, really starting in 2020. I believe there are various forms of wellness—Physical, mental, emotional, and spiritual—and you can find it in different mediums.

If you can find leaders that provide all that I mentioned prior AND also support your wellness, these are precious gems and may be rare. Even better is if you become one of those leaders and help make it the norm.

I can confidently say that putting a focus on wellness over the past few years has kept me resilient through some tough activities like global supply challenges.

I would be happy to share with anyone the realness of my wellness journey. It is a journey, not a destination, and takes a lot of time, work and care. We can continue the wellness conversation on my Instagram handle: @peace.of.kak

Take care,

Kirstin K. | Technology & Operations Leader for business outcomes and people

Letter 5

Dear reader,

I would like to share with you a short list of best practices that have helped me and others that I have coached or mentored in the past to advance in our careers:

1. *The first one and most important is that you have to take control and own your career development. This is something that can't be delegated, and you must be the full owner of your career progression.*
2. *It is also critical, as you define what you want to do next and when new opportunities develop for you, as part of your decision-making, consider whether the opportunity that you are considering next is the best alternative to help to get you closer to where you want to go next. Most people*

don't think that long-term, and they could be attracted by what might look like a good short-term opportunity that may take you away from the end goal, which is where you want to go.

There are some foundational building blocks that you need to prioritize as you work in your career development and progression:

1. *This is the era of specialization, and knowledge is everything. It is then important that you continue to prioritize acquiring new skills, especially those that will better position you in the areas you aspire to grow or advance.*
2. *The next one sounds like a cliché, but it is a foundational block that you can't ignore. It is important as you develop your career plan that you break it into phases or smaller pieces but with clearly identified goals you need to achieve. All of them, accumulative, will eventually get you where you need to go and will keep you focused on the important things that will help you to advance.*
3. *Be flexible with the timeline of your targets since career development is a long journey, and you need to be patient but remaining focused on the end goal. If certain milestones don't necessarily happen at the exact target you had originally planned, this is not as critical as long as you keep going in the right direction. If you don't do it, you may get frustrated if the completion of your key milestones doesn't happen exactly when you originally targeted them. On the opposite side, there are some that may be accomplished earlier so things will average down. The key is to continue to follow your North Star!*
4. *It is important that you realize that you will not accomplish everything you want to do by yourself, so it is critical that you develop early in your career team effort and collaborative skills. Being surrounded by others that can help*

you to grow and accelerate your career is a key component of the tool kit you need to develop. People that have strong networks are more likely to succeed than those that don't. Plus, being a team player will help you as you move into leadership roles.

5. *Even if you do all of the above but you don't deliver results, it may not matter, so it is crucial that you understand that the catalyst and the key to your sustained career growth is to consistently deliver good results and meet your commitments. This will give you a level of credibility and dependability that will open many doors for you.*
6. *Taking control of your career means not just the early planning and execution of it, but you also need to take control of being your own advocate, and the best way to do it is to speak up. If you don't do it, others will not do it for you. Being able to clearly state and communicate what you do and why it is important and doing it with confidence is an important foundational block and skill that you need to include in your career development plan. The more and the better you do it, you will not only communicate your messages, but you will show your capabilities based on the confidence you show as you share with others the output of your work and contributions.*
7. *The best job that you will ever get is a job that leverages your core competencies, and something you are also passionate about. The end result is going to be outstanding performance and contributions, and success will bring more success, and you will advance in your career, so don't make the mistake to work on something you don't like and you are not passionate about. It will take you away from where you need to be. Focusing on your strengths and working on things you like and are good at is the best formula of success.*

8. *Within the umbrella of not being able to do everything by yourself, it is important that you leverage as early as possible in your career a strong network of mentors, coaches, and sponsors. Each one of them will catapult you in your career progression.*
9. *Be ready for opportunities, since they may come at any time. Volunteer if any opportunity of interest to you becomes available. Be willing to stretch yourself and take challenges even if they take you away from your comfort zone. Challenging yourself and taking new opportunities will keep you sharp and give you the opportunity to learn new skills and/or discover new capabilities you did not know you had before.*

Hopefully, you find these simple tips useful as you start your career development journey!

Best regards

Mario Bolanos | Senior Director, Worldwide Regional Applications Engineering at Intel Corporation. Latino Corporate Directors Association (LCDA), HITEC100, SHPE Star Award, McKinsey Executive Leadership, Hudson Institute Licensed Coach

Letter 6

Dear Reader,

We often hear about the "role-required skills" and other soft skills or "nice to haves." My advice is find what makes you unique and spend equal time building the necessary tactical skills to be successful in the role you're seeking, as well as the interpersonal, communication, and leadership skills that will accelerate your personal and professional development. These so-called soft skills are some of the most crucial that will set you apart from your peers. Early on, identify a mentor or coach that can help you build your short and long-term goals. Creating a network and community of people creates a safe place for you to reskill and upskill in addition to your manager. You gain a better understanding of different perspectives and learn about the variety of leadership styles that exist to help you truly find yours. And when you do… pay it forward and start mentoring others! Not sure where to start? Be curious. Curiosity leads to continuous learning and will be a catalyst for you to identify and achieve your goals. In today's remote/hybrid work culture, it can be fairly easy to feel isolated or be heads-down doing your individual contributor work. My advice is to take 15 to 20 minutes a week to connect with folks outside of your team and build relationships across your organization. It may feel minuscule, but sometimes it's those spontaneous discussions that lead to the greatest ideas!

Sincerely,

Preeya Patel | Vice President, People

Letter 7

Dear Reader,

Based on the immensely positive impact it has had on my career, I would strongly recommend taking advantage of opportunities to continue learning new skills via academic courses as well as professional certifications, especially for young and early mid-career professionals.

Having skills in a complementary domain and being able to demonstrate how to apply those to build better solutions to problems in your main domain of work can get you hired for roles where the preferred qualifications and experience may have been higher than yours. Learning usable techniques from relatively newer domains also helps avoid your skillset being seen as outdated, especially if your area of work undergoes a paradigm shift as technologies evolve. Furthermore, having skills in multiple domains can help in finding employment when particular industry segments where your core skills are most valued go through an economic downturn resulting in job losses, but other sectors that are interested in your complementary skills are still hiring.

Also, especially if your job role is more on the research side, do not discount the value of professional certifications in your own field of work. In industry, the chances are that your solutions can only have a measurable positive impact on the business if they get deployed. Having insights into how solutions can be deployed, and how folks on the operational side of things think, will both help you invent solutions that are deployable and help you bridge the communications gap that often exists between research and operations.

Moreover, most companies have an education reimbursement policy, so not taking advantage of it is really leaving a very valuable benefit on the table!

Ramesh Subbaraman | Technology Leader

Letter 8

Dear Reader,

As someone who has gained valuable experience during my career, I want to share some advice with you as you embark on your journey. These insights, influenced by my own experiences and wisdom, will help you navigate your journey and path to success:

Embrace a learning mindset. Never stop learning. Be open to new ideas, experiences, and knowledge. Seek out mentors; continuously improve your skills. The willingness to grow and adapt is key to long-term success.

Building strong relationships. Foster positive relationships with your colleagues. Be empathetic, respectful, and willing to collaborate.

Communication is at the heart of any successful endeavor. Express your thoughts clearly and provide feedback. Listening is the key to successful communication. Be a listener.

Take ownership and be truthful. Making mistakes is not failure; it's important for growth. Not being trustworthy and transparent is career-limiting. Stay persistent.

Maintain work-life balance. While it's important to be dedicated, it's equally important to maintain a healthy work-life balance. A balanced life helps you perform your best.

Kathye Mendez

Letter 9

Dear Reader,

If I were to wind back the clock to 19 years ago when I started my professional career after college, there are five things that I would tell myself to do better at:

- *Build relationships – Work is about people. How we treat people every day is a choice. Every day we make hundreds of choices that define how we interact with each other, and the bearings the interactions have on our future work. This might sound clichéd but in summary: Treat the people the way you want to be treated, follow people the way you want to be followed, and lead the people the way you want to be led.*
- *Constantly seek feedback – Consistently and regularly seek feedback on how much value you add to your team and*

organization. Ask for feedback from your peers, your manager/s, your direct reports (if applicable), always reminding them to be candid and open. Listen actively to the feedback. It is okay if some of it is critical or even probably disappointing. It is better to be disappointed rather than be shocked during the year-end review. The way to grow is to analyze and act on critical, constructive feedback.

- *Communicate effectively – Make effective use of both your body language and verbal communication. The key to effective communication, especially in situations that involve conflict, is effective and active listening: for instance, ask simple questions (how/ why), then listen patiently. This will help you respond instead of simply reacting. I believe that communication when done effectively leaves all parties satisfied, respected, and is the best form of self-expression, leading to building healthy relationships (going back to the first bullet).*
- *Build Credibility - "Do what you say that you will do," no matter if it's hard or unpleasant. This shows that you are credible, sincere, truthful, all of which are building blocks of a good, solid, personal brand.*
- *There is no "one-size-fits-all" approach for career success. In other words there is no 'blueprint' for career success that works for all. Stay open to analyzing alternative ideas and viewpoints. Most people have a range of paths to achieve career success. To discover these paths, I suggest spending 10 percent of your waking hours (or whatever time that makes sense to you) to explore something that catches your interest (could be cryptocurrency, reading about AI, playing an instrument etc.). This not only helps one to take one's mind off work but allows one to cross-pollinate ideas from other avenues and to increase creativity.*

Hindsight is 20-20, but as I approach the midpoint of my career, these five learnings from my professional life are ones that I believe will help me position myself for more success in the next half of my career. Thank you for reading and hope you find this relevant and useful.

Vidya Anandan | VP of Strategy at First Republic Wealth Management

Letter 10

Dear Reader,

If you want to propel yourself in your life and your career, you need to first pull yourself back.

Think of a bow and arrow. You need to first pull the arrow back to create the momentum. Then, once you let go, it flies forward.

The pulling back, for you, would mean introspection, self-awareness, and mastery of your own emotions.

Once you can master your 'self,' you are in control of your presence and your charisma. This defines you. This becomes your superpower.

When your superpower harmonizes with your interactions—your presence becomes a force that inspires, motivates, and elevates those around you.

Mastering your 'self' allows you to be your most authentic self. Authenticity is the most powerful frequency humans can emit. People are drawn to authenticity; it forms the bedrock of trust and rapport in relationships, whether personal or professional.

Authenticity empowers you to navigate life with conviction and purpose. It gives you the courage to stand firmly in your be-

liefs, express your ideas unabashedly, and embrace your uniqueness.

By being your most authentic self, you invite others to do the same, fostering an environment where everyone feels empowered to show up as their genuine selves.

Reading this book is an excellent step towards building your self-awareness and growing into your authenticity.

I applaud you for investing in your future self. There is nothing more valuable and nothing more worth your while.

Remember, your authenticity is not just your power; it's the catalyst for transformative connections and profound change.

Dare to be authentic; the world craves your genuine presence.

Sincerely,

Fizzah Bajwa | Writer, Coach, and Podcaster at Grounow

Letter 11

Dear Reader,

What does it mean to be a leader?

Leaders sell a vision.

Leaders have to inspire and persuade—to influence direction towards a vision, be that technical, business, or market. They have to sell why it is important, the options to get there, and why a path is chosen. There's a negative connotation around sales, but every one of us sells: it's part of everyone's job, just some might be selling hardware products, some software programs, and some visions.

There is an art to it: it's the art of persuasion—inspiring, getting people to view the benefits for them in the largest sense, personally and beyond. It needs to be practiced and honed. And it needs to be adapted according to the team and the situation.

Establish key principles.

Create a set of shared guiding values: an ethos for yourself and others. Contract these with your team and stakeholders and demonstrate using them to hold yourself and others to account, with the agreement that they do the same. The test of leadership is in times of stress can you live up to the values—and use them to convince others to come with you.

Communicate, listen, & learn.

Communication is not just one way—you need to continually learn from others and to test your ideas. Your own vision isn't always right—so test it out. In business and in leadership, there's more than one right answer and more than one path to get there: it's much more about willingness to think through the paths. And when you communicate the path, take time to explain and bring people with you.

Share successes, own failures.

If things go well, look out across your team and collaborators—that's where the success came from. When things don't go well, you look in the mirror. That's where failure lies.

A leader works on the system, not in the system.

Leaders don't design, or do the hard creative work. They create a system in which the organization can do that successfully. The most common misstep is the failure to transition from doing it yourself to enabling others—to make the system as a whole more effective. In that transition to leader, you need to be adaptable: to understand where to spend time, how to prioritise, and how to add value.

Most of all, have fun doing it.

Think carefully about your career objectives. What is your 'Ikigai'—the overlap of things you love, are good at, that you can be paid for, and what the world needs? Maybe you have found that already.

Else, what does success look like for you? Remuneration, recognition, responsibility? If you have your Ikigai goal, take an inventory of your skills—your strengths and weaknesses—and plan your progress towards it. What's missing? What do you need to strengthen? How will you do this? Mentoring, training, assignments? Realise there are steps in the journey, and some will be sideways or even backwards.

One last point. Everyone is a leader. We are all leaders of our own behaviours and of the influence and impact we have on those around us. All the points above apply, whether you're a CEO leading a whole organization, or a leader just starting out on your career.

Mark Jervis | Product Planning Leader

Letter 12

Dear Emerging and Seasoned Professionals,

I want to share some of the key lessons I've learned in my ~ 2 decades of working in the tech industry, spanning from a small company to midsize to an established tech giant:

1. *It's great to be passionate about a thing and pursue that with full determination and grit. But also, it's okay if you haven't figured out what that passion is. Even if you are 80 years old and still trying to figure that out, that's okay.*
2. *Please remember that it's your unique life and all your unique experiences. And surely your career must've had moments of accomplishment and passionate pursuits. So please be sure to reflect and journal as much as you can in whatever format works best for you. You could take walks out in the sun and dictate your experiences into the phone. You could type it up and save it in your personal device, blog it, vlog it, or even take the good old pen to paper.*
3. *Whatever job it is that you are in, please wear a mindset of 'I am going to do the best job I can, but without compromising my health, family and peace.' And keep exploring different roles, especially when opportunities come knocking your way. Give it a shot so you will know whether you like it, love it, hate it, or can live with it. It might seem scary to dive into things without knowing a lot about the domain or job. But the human mind can truly achieve things with a learning mindset. It's absolutely okay to dive in and learn on the go.*
4. *It's so vital to find a safe tribe no matter which job you take up or which environment you are in. This safe tribe would be people you come across, who value you more than you do yourself and are willing to make time for you, advocate for you, and sponsor you. Or maybe even just be there to listen to you without judgment and provide support. It's not easy for everyone to figure out who the right people are to trust and not to, but their actions will always speak louder than words. I believe in giving people multiple chances until they show a strong reason not to.*

And I've always trusted my gut and instinct and it's never let me down.

5. *In any job, it's important to pay attention to the key decision-makers, visible and invisible influencers and manage these stakeholders by working well with each in a customized way. Learning how to work well with people and manage stakeholders is not rocket science and can be learned. It's so important to invest in a really good career coach who is not just a coach but also your sparring partner for your career, someone who acts and thinks like your clone but with your best interest at heart. You can practice stakeholder management, communications, confidence, and strategizing the best way to showcase yourself and your brand at work, by collaborating very closely with a career coach.*

Having said all this, I want to emphasize that your job is just a part of your life. Please don't make it your identity and life. No job is worth your mental and physical health.

Keep rocking, keep being you, and you are good enough as you are.

Sincerely,

Mahathi | Technologist, Business Innovation, Sales and Customer Experience

Letter 13

Logical Levels[44]

Hurry Up Slowly can be of tremendous help for career development at many levels. To explain what I mean by this and, I hope, add a little more value also, I'll give a quick introduction to the concept of Logical Levels.

Logical Levels allow us to stand back from complex situations and look at them from alternative viewpoints. This ability is tremendously useful with difficult, career-impacting decisions, for example.

I run ICONDA Solutions, a multifaceted business which provides coaching and training to companies all over the world and is also active in software sales. Having integrated Logical Levels into my personal toolkit some time ago, I use it to regularly understand how I should shape a commercial discussion. I also use it when coaching and training to quickly identify my clients' most important issues. I am sure that you can apply it to your own role and additionally, in almost any context, use it as a way to step back from a conversation and assess the big picture.

To solve a problem, I have to attack it at the right level. If I am debugging software, I don't try to fix a networking issue by immediately looking at the assembler code of my computer's OS. Likewise, when dealing with individuals, teams, and organizations, it is important to know where to start and in what order to proceed. Just as computer systems are layered (operating system, utility, application, etc.), so there are logical levels for dealing with human systems:

Identity	Family figure, Thought Leader, Painter, Athlete, Customer-Facing Engineer, Comic, ...
Values, Beliefs	Trust, honesty, authenticity, passion, resilience, conscientiousness, ...
Skills, Knowledge	Planning, encounter management, discovery, influencing & challenging, coping with complexity, ...
Behaviours	Use of expert knowledge, proactivity, preparedness, resourceful, teamwork, excellent comms, on time, ...
Environment	Clients, projects, technology, etc.

Take a moment to think about one or two situations that you are dealing with at the moment. Assuming that you have limited influence over the environment, is the central issue at the level of what to do and when—the behavioral level? Or is it more to do with how to do something—do you have the necessary knowledge and skills, and, if not, how can you acquire them? Maybe the core problem is at an even higher logical level and concerns your values and beliefs? This would be the case if you had to decide between staying in a secure, well-paid position, working with people that you liked, or taking a risk and doing something that looked incredibly interesting. Some decisions even take us to the level of our identity—who do you want to be?

I hope you find this to be nurturing food for thought and wish you well in your career.

Andy Betts | CEO and Founder at ICONDA Solutions | https://iconda.solutions

Letter 14

Dear Readers,

It is my honor and privilege to write to you all. It is said that "Investing in yourself is the best investment you will ever make." I think you all have made that good decision by investing time in reading this well-compiled book.

My top three ideas to share with you all:

1. *Life is too short. Have no regrets. Follow your dream; listen to your heart.*
2. *Be ruthless with your prioritization. Crystal clear with respect to P0 vs P1. Know that P1 can be your work, and it can wait. P0 can be family, especially in these trying times.*
3. *Take care of yourself; you can take care of the rest if you are sane. Follow the airplane oxygen air mask principle.*

Slowing down in life: *Life is not all about numbers, although humankind is fascinated by the fact that our minds can steer us from ruin if we control the little things in life. The human mind, by its very nature, represents a mutable state. We change. We grow. Hence our love to trace this beautiful trajectory. We count the number of hours we sleep, number of miles we run, the number of steps we take every day, amount of calories intake, distances driven to and from, and whatnot. We obsess over the lines of code written, the number of open customer issues, number of build failures, open/filled headcounts, number of wins/losses, customer adoption rate, and performance metrics. Let's slow down and take a p-a-u-s-e. Let's count the number of moments that take our breath away! Thankful for being alive to experience all the tiny precious events—the first laugh on my babies' beautiful faces, the feel of raindrops on my cheeks and wind in my hair, lungs full of clean air, the vibrant colors of the butterfly, the sight of birds flying in the clear blue sky, the whirring wind and the humming bees. Life is and will be alright.*

Prioritization: *The last couple of years have tested people's resilience and grit. I am amazed to see the lines between work and life blurring. That has stressed a lot of folks. Here is my attempt to paint a take on a tangential perspective. It took a pandemic for corporations and families to realize that work and life indeed go hand in hand. At times, work takes a priority, and family takes a backseat. There are times when family needs your focus, and work has to wait. You can't help but shirk your responsibility and escape a little. Now, more than ever, you need a supportive network of friends and family at home to succeed at work. You need an empathetic and resilient team at work for you to maintain sanity at home. I would like to believe that I have always taken a keen interest in my teams, and the last couple of years has strengthened that bond way further. My cube mate these days is my five-year-old who partakes in many of my 1:1 conversations. A couple of* *months back, she sensed a rather crucial conversation with one of my coworkers. To ease things, she jumps into my Zoom call and says, 'Please tell him that my Barbie says Hi.' And we both blurt out laughing. The other day, she almost noticed a tear when one of my peers was pouring her heart out. My cube mate comes in with, 'Can you share that I loved the Ballet background she had the other day, and I have some new Ballet moves too?' My team loves these tiny interruptions, and I love theirs. I now appreciate not just my team members' spouses and their kids, but their family passions and interests too.*

Take care of yourself: *You are you-nique. Be you, always. Bring your authentic self to every place you go. The world needs the real you and the whole of you. Share your passions with people around you. Be there to inspire and be inspired. Pursue hobbies; this is a great way to shake off that perfectionist inside you. As an example, I like to run but am far from the best runner. However, I have realized that my best runs are not my fastest or longest, but they are the ones where I had the most fun. They may have been slow and/or short, but I enjoyed each moment of every stride, cherishing every split second. If I were to counsel anyone in the job market right now, my advice would be to pick a job that you will have fun working at, a role that you enjoy and with folks that you like being around. Period. Longevity, career progression, title advancements, and compensation are all important, but the fun aspect trumps it all.*

Let me share a little bit about myself. I grew up in a traditional South Indian joint family household in Mumbai, India. We lived a hand-to-mouth existence, but my parents showered so much love and care that we as kids always felt rich. I will always cherish my mother's advice to never forget my roots and remember where I came from. If I could write to my younger self, I would thank her.

- *Thank you for pursuing the engineering career that your mother wanted for you! Someone in the family once told her 'She is tall, dark, and now engineering? Seriously, how will you ever find a guy for her?' I chuckle every time I think about it now.*
- *Thank you for disregarding the "advice" from one of my previous managers who I highly respect. He counseled me to take a five-year break so I could raise my family. I am proud today to be a role model to my two little girls, who know very well that they can work and have a life too if they like.*
- *Thank you for following your heart and choosing the partner that loves and respects you! Thanks for fighting the stigma and initial setbacks. Little did I know back then that this would be one of the most important decisions shaping my future career.*
- *I am passionate about developing the next set of women leaders. I recently joined a couple of women in engineering communities in the Bay Area, SheTO and the NewClub. Getting a chance to meet incredible peers in unique settings has been so valuable. This mission hits home close since I have two little girls who are watching and learning from my every move and action. I would love to see more gender balance by the time they join the workforce. For them, for my mom and other women, I believe it is worth striving for a more equitable and diverse work culture. To all the young ladies out there, go ahead and pave the path for more women. Break the mold for what an engineering leader should look, sound, and dress like! Connect with me on LinkedIn: https://www.linkedin.com/in/sharadha/*

Sharadha Venketasubramanian | Software engineering executive leader

Letter 15

Dear Reader,

I want to take a moment to share two important aspects that I have learned in my life—the significance of faster decision-making and the importance of being kind to yourself.

In our fast-paced world, the ability to make timely decisions is a skill that can greatly impact our personal and professional lives. Decisions, whether big or small, shape our future. The sooner we make a decision, the sooner we can adapt and progress. Procrastination often leads to missed opportunities and unnecessary stress. While it's essential to gather information and consider options, there comes a point where we must trust our instincts and move forward. Embrace the power of decisive action, for it can lead to growth, learning, and the achievement of your goals.

Equally important is the practice of being kind to oneself. In the pursuit of success, we can sometimes be our harshest critics. We set high standards and demand perfection, often forgetting that we are only human. Remember that mistakes are growth opportunities, and self-compassion is the foundation of resilience. Treat yourself with the same kindness and understanding that you would offer to a dear friend. Embrace your imperfections, celebrate your achievements, and nurture your well-being. By doing so, you not only enhance your own life but also become a source of positivity and support for those around you.

As you delve into the pages of this book, I encourage you to keep these two principles in mind.

May your journey be guided by both wisdom and self-compassion, and may you find inspiration and insights within these pages to enrich your life.

With warm regards,

JP Singh | Strategic Marketer & Data Analyst Leader at Intel Corporation

Letter 16

Dear Readers,

Drawing upon the valuable experiences I gained throughout my career, I aim to offer guidance for your journey. Rooted in personal insights and wisdom, these pieces of advice are intended to assist you in navigating your path toward success.

To enhance efficiency in the workplace, it is crucial to prioritize tasks and manage time effectively. Begin by creating a detailed to-do list, categorizing tasks based on urgency and importance. This will help you focus on high-priority items and avoid feeling overwhelmed by a lengthy list. Employ project management tools like ClickUp to organize your tasks. Break down larger projects into smaller, more achievable tasks and establish practical deadlines for each segment.

Furthermore, fostering effective communication and collaboration is key to optimizing workflow. Regularly communicate with team members to stay informed about project updates, deadlines, and potential challenges. We also embrace ClickUp as our collaborative tool to streamline communication and ensure everyone is on the same page. Delegate tasks according to individual strengths and expertise, promoting a sense of ownership and accountability within the team. Regularly review and adjust your work processes, identifying areas for improvement and implementing changes to enhance overall efficiency. By fostering a collaborative and organized work environment, you can maximize productivity and achieve optimal results.

Angeli Acapulco | Digital Marketing Specialist at Grounow

Conclusion

As we close this transformative journey through *Hurry Up Slowly*, it's time to reflect on the lessons learned and the steps ahead.

You began this journey seeking clarity, direction, and wisdom to navigate your career. You discovered your purpose—that compelling compass guiding your decisions, actions, and growth. You harnessed knowledge as your tool, understanding that continuous learning is the key to innovation, evolution, and success.

You faced challenges head-on, recognizing them as opportunities for growth, rather than roadblocks. You embraced tension and conflict, knowing they are springboards to resilience, poise, and effective problem-solving. You set your sights on a clear vision of your desired destination, building essential skills and habits to propel you toward it.

Throughout this journey, you learned the importance of consistency and intentionality in career development. You saw how effective time management, networking, and self-care practices become the pillars upon which you build your success.

In the resolution, you realized that growth is an ongoing journey, not a destination. You learned to adapt to new challenges and opportunities while staying true to your evolving values. You embraced lifelong learning, ready to seize every opportunity that comes your way.

Finally, you soaked in the collective wisdom of industry experts who traveled similar paths. Their insights and perspectives illuminated your path, accelerating your growth and empowering your career development.

Standing at the precipice of a future filled with purpose, growth, and fulfillment, you know that your career development adventure is about more than just reading—it's about engaging with the exercises and lessons provided actively.

So, dear reader, remember that your journey does not end here as you close this book. Act, invest in yourself, seek professional help, and watch your potential unfold. Identify areas for growth, act toward your goals, and build a supportive network. Embrace patience and resilience, and approach your career development intentionally. Remember, success is not a destination but an ongoing process; the true joy lies in the journey.

As you embark on this exciting path, let *Hurry Up Slowly* be your guide, inspiring you to push beyond your limits and create a fulfilling and meaningful future. Your adventure awaits.

Become the leader you wish you had!

SHARING KNOWLEDGE: THE FINAL STEP

As we close this book, there's one more lesson I would like to share. It revolves around the power of sharing itself. Like a candle flame, knowledge does not diminish when used to light another candle. Instead, it spreads light farther, illuminating the path for others just as it did for you. I humbly request that you share the insights you gained from *Hurry Up Slowly* with those around you.

The book's title, *Hurry Up Slowly*, carries within it a profound truth. In our rush to get ahead, we often forget the importance of slowing down and taking the time to learn, reflect, and grow. As you journeyed through these pages, I hope you discovered that to hurry up, one must first learn to slow down.

In your hands, you hold a book and a torch passed down from those who have walked the path before you. Now, it's your turn to pass it on, spread the light, and help others navigate their paths in the tech industry.

Remember, true success is not measured by what you gain but by what you give. Sharing knowledge is one of the greatest gifts you can offer, enriching not just others' lives but your own as well.

So, as you step into the next chapter of your professional journey, remember to slow down, learn, grow, and, most importantly, share. Here's to your journey, your growth, and the extraordinary career that awaits you. Your next chapter begins now.

Thank you for joining me on this journey. I look forward to hearing your stories of success, learning, and growth. Until then, keep learning, keep sharing, and remember—to hurry up, you must first learn to slow down.

Endnotes

1. Fe del Mundo. (2023, November 18). In *Wikipedia*. https://en.wikipedia.org/wiki/Fe_del_Mundo.
2. Masaru Ibuka. (2023, November 26). In *Wikipedia*. https://en.wikipedia.org/wiki/Masaru_Ibuka.
3. Radia Perlman. (2023, October 19). In *Wikipedia*. https://en.wikipedia.org/wiki/Radia_Perlman.
4. Mary Jackson (engineer). (2023, November 18). In Wikipedia. https://en.wikipedia.org/wiki/Mary_Jackson_(engineer).
5. Brewster Kahle. (2023, November 15). In *Wikipedia*. https://en.wikipedia.org/wiki/Brewster_Kahle.
6. Bryan Stevenson. (2023, October 7). In *Wikipedia*. https://en.wikipedia.org/wiki/Bryan_Stevenson.
7. Nadine Burke Harris. (2023, November 10). In *Wikipedia*. https://en.wikipedia.org/wiki/Nadine_Burke_Harris.
8. DeRolf, S. (1997). The Crayon Box that Talked. Random House Children's Books.
9. Pluto. (2023, December 1). In *Wikipedia*. https://en.wikipedia.org/wiki/Pluto.
10. Alfred Wegener. (2023, November 19). In *Wikipedia*. https://en.wikipedia.org/wiki/Alfred_Wegener.
11. William Shakespeare. (2023, December 1). In *Wikipedia*. https://en.wikipedia.org/wiki/William_Shakespeare.

12. Ursiny, T. (2003). The Coward's Guide to Conflict: Empowering Solutions for Those Who Would Rather Run Than Fight. Sourcebooks.
13. Thomas–Kilmann Conflict Mode Instrument. (2023, July 1). In *Wikipedia.* https://en.wikipedia.org/wiki/Thomas%E2%80%93Kilmann_Conflict_Mode_Instrument.
14. Maslow's hierarchy of needs. (2023, December 3). In *Wikipedia.* https://en.wikipedia.org/wiki/Maslow%27s_hierarchy_of_needs.
15. University of Warwick. (n.d.). New study shows we work harder when we are happy. Retrieved December 6, 2023, from https://warwick.ac.uk/newsandevents/pressreleases/new_study_shows/.
16. Navarra, K. (2022, April). The Real Costs of Recruitment. SHRM Online. https://www.shrm.org/resourcesandtools/hr-topics/talent-acquisition/pages/the-real-costs-of-recruitment.aspx.
17. Gallup, Inc. (n.d.). The Right Culture: Not Just About Employee Satisfaction. Retrieved December 6, 2023, from https://www.gallup.com/workplace/231602/right-culture-not-employee-satisfaction.aspx.
18. American Psychological Association. (n.d.). Journal of Applied Psychology. Retrieved December 6, 2023, from https://www.apa.org/pubs/journals/apl.
19. National Institute of Mental Health. (n.d.). Mental Illness. Retrieved December 6, 2023, from https://www.nimh.nih.gov/health/statistics/mental-illness.
20. Betts, A. (2015). Client Encounters of the Technical Kind. Iconda Publishing.
21. Art of asking questions - Betts, A. & Ethirajan, S (2023, February 3). How asking questions can help you achieve your goals. Retrieved December 6, 2023, from https://www.linkedin.com/pulse/how-asking-questions-can-help-you-achieve-your-goals-grounow-1c/.
22. United Nations. (n.d.). The United Nations Standard Minimum Rules for the Treatment of Prisoners (the Nelson Mandela Rules). Retrieved December 6, 2023, from https://www.un.org/en/events/mandeladay/mandela_rules.shtml.
23. Kumar, A. (2018, November 5). Dropbox's 'Aha!' Moment: Unleashing Explosive Growth Through Product Analytics.

LinkedIn Pulse. Retrieved December 6, 2023, from https://www.linkedin.com/pulse/dropboxs-aha-moment-unleashing-explosive-growth-through/.

24. CBS News. (n.d.). Why Netflix users cut the cord: Cable isn't worth what it wants. Retrieved December 6, 2023, from https://www.cbsnews.com/news/why-netflix-users-cut-the-cord-cable-isnt-worth-what-it-wants/.
25. Richard Branson. (2023, December 3). In Wikipedia. https://en.wikipedia.org/wiki/Richard_Branson.
26. Steve Jobs. (2023, December 2). In Wikipedia. https://en.wikipedia.org/wiki/Steve_Jobs.
27. Slack (software). (2023, December 6). In Wikipedia. https://en.wikipedia.org/wiki/Slack_(software).
28. Sundar Pichai. (2023, November 24). In Wikipedia. https://en.wikipedia.org/wiki/Sundar_Pichai.
29. IBM. (2023, December 4). In Wikipedia. https://en.wikipedia.org/wiki/IBM.
30. Jeff Bezos. (2023, November 28). In Wikipedia. https://en.wikipedia.org/wiki/Jeff_Bezos.
31. Rohn, J. (n.d.). Either you run the day or the day runs you. Retrieved December 6, 2023, from multiple sources.
32. Practice in Public. (2023). 7 Pieces of Timeless Success Advice from Jim Rohn. Medium. Retrieved December 6, 2023, from https://medium.com/practice-in-public/7-pieces-of-timeless-success-advice-from-jim-rohn-cf1de166f844.
33. Huda Kattan. (2023, November 28). In Wikipedia. https://en.wikipedia.org/wiki/Huda_Kattan.
34. Cristina Junqueira. (2023, December 5). In *Wikipedia*. https://en.wikipedia.org/wiki/Cristina_Junqueira.
35. Lourenço, D. (n.d.). Tech workers beware: 5 'good habits' that are actually hurting your career. Retrieved December 6, 2023, from https://www.linkedin.com/pulse/tech-workers-beware-5-good-habits-actually-hurting-your-louren%C3%A7o/.
36. Paymo Team. (n.d.). 30 Good work habits in the workplace to boost your career. Retrieved December 6, 2023, from https://www.paymoapp.com/blog/work-habits/.

37. Bit.ai Team. (n.d.). 19 Work habits to get the career growth you deserve! Retrieved December 6, 2023, from https://blog.bit.ai/work-habits/.

38. Puckett, C. (2022). Redefining Geek: Bias and the five hidden habits of tech-savvy teens. University of Chicago Press.

39. Indeed Editorial Team. (n.d.). 14 Bad habits at work (And how to improve them). Retrieved December 6, 2023, from https://www.indeed.com/career-advice/career-development/bad-habits-at-work.

40. Semuels, A. (2023, June 14). You're not imagining it—job hunting is getting worse. Time. Retrieved December 6, 2023, from https://time.com/6287012/why-finding-job-is-difficult/.

41. Abisoye Ajayi-Akinfolarin. (2023, November 3). In Wikipedia. https://en.wikipedia.org/wiki/Abisoye_Ajayi-Akinfolarin.

42. Coco Chanel. (2023, November 29). In Wikipedia. https://en.wikipedia.org/wiki/Coco_Chanel.

43. Yves Saint Laurent (designer). (2023, November 27). In Wikipedia. https://en.wikipedia.org/wiki/Yves_Saint_Laurent_(designer).

44. Dilts, R. B. (2018). Changing Belief Systems With NLP. Dilts Strategy Group.

Acknowledgments

My most immense gratitude goes out to my wife (Sharadha), my children (Diya & Zara), my mother (Girija), and my sister (Renu). My existence, love, joy, and intelligence are tied significantly to them.

Next, I want to thank my in-laws Saroja and Subramanian, who introduced and taught me the meaning of Hurry Up Slowly.

I'm fortunate to have the love and support of friends. I can't imagine how I can return the favor to them. At the pinnacle are these friends from different worlds—Josh, Ramesh, Amit, and Ashu. They have been a beacon during my darkest times.

The following friends of mine have made my life rich beyond comprehension. I'm forever grateful for them—Priti, Austin, Rina, Sampath, Rich, Ghazaleh, Marion, Todd, Mike L, Yi-Hui, Anna, Mike G, Stephanie, Mickie, James, Nandini, Andy, Jason, Basil, Prashant, Varun, Tanmoy, Staci, Nupur, Mark, Susannah, Gloria, Marlon, Steve S, Steve E, Ray, Neil, Uriri, Karl, Anuj, Sanjay, Prathibha, Vivek, Neha, Shilpa, Raja, Bindu, Anukriti, and Aparna.

Shout-out to friends who have made Grounow successful—Angeli, Jack, Chriszel, Samarth, Andy, Fizzah, Kirsten, Jeanne, Trevor, Sam, Darwin, Breven, Dinah, Megane, Ryan, Kirstin, JP, Ellie, and Jonathan.

To all my teachers, and here are the ones I want to call out especially—Mrs. Beena, Mrs. Juliet, Mrs. Merlin, Mrs. Parvathi, Mrs. Rajeshwari, and Mrs. Radhika. Thank you for your guidance and wisdom.

My heartfelt appreciation goes to my extended family—Siddhanth, Vedanth, Ishanvi, Smita, Shankar, Venkat, and Avi—and my grandparents, A.S. Rajan and Vijaya Kumari, my Patti and Vishnu Priya.

I'm thankful to you for picking up this book and utilizing it!

Made in the USA
Monee, IL
04 November 2026

67379721R00216